BRIDGING EXPECTATIONS

Healing the Divide Between Those Who Give and Those Who Expect

Dr. Eric Edokpaigbe Idehen

BRIDGING EXPECTATIONS

Published by:
Eleviv Publishing Group
Centerville, OH 45458
info@elevivpublishing.com
www.elevivpublishing.com

ISBN: 978-1-7331416-1-1 (PB)
        978-1-952744-92-1 (eBook)

Printed in the United States of America

"Being a human being is to be human. I have always been curious why the diasporans sometimes have issues helping their folks in their native land. I did not understand why. Now I know!

TABLE OF CONTENTS

DEDICATION

This book is dedicated to the diasporan—who, despite the distance, continues to give, love, and show up for home, even when the burden is heavy.

*You may not always see the results of your kindness, but every bit of positive energy you contribute to the world makes it a better place for all of us. -**Lisa Currie***

INTRODUCTION

Between the 1950s and 1980s, many Nigerians left the country in search of better educational opportunities in the West. Many received scholarships to study abroad. Back then, family expectations from back home were minimal or non-existent. During this time, the expectations from family members back home were minimal. Often, families in Nigeria sent money to support their loved ones studying abroad.

Most of these individuals returned home after completing their studies to contribute to the development of their country. The 1980s saw a new wave of Nigerians leaving the country, commonly referred to as the ***"Andrew's wave" or the "I'm checking out" syndrome***. However, in the early 1990s, there was an increase in emigration, followed by an unprecedented exodus in the late 1990s and mid-2000s.

Individuals were pursuing better opportunities. The primary motivation for this exodus was the pursuit of a better life, driven by economic reasons. People left for work, education, or both. I belong to this group, along with many others from my generation.

There was also another peak migration in the 2000s. That was when many Nigerians took road trips through Morocco to reach Spain or Libya, and then on to Italy. Again, the reason was to pursue better economic opportunities, which continued for nearly two decades. Between 2015 and 2023, another wave of Nigerians migrated to seek a better life, with skilled workers and students emerging as a new phenomenon—the 'Japa' wave. For the first time, all the economic classes of Nigerians, from the affluent to the extremely poor, participated in the migration. As the economy was in decline, the primary motivation for the diaspora was to support their own families back home. Anything would be done at any cost to help out, even if it meant availing of loans via credit cards. Therefore, the people who were settled in rural regions with agricultural activities shifted to city locations and maintained a luxurious life, staying in duplex houses and possessing multiple cars, similar to those in Belgium.

Such people were even bestowed with electronic appliances for status enhancement. In addition, a few of the parents became members of high-class social organizations, attending events in a flashy style and contributing money liberally. As time passed, families became increasingly dependent on financial support from those in the diaspora, yet were less productive than they had been.

In the early 2000s, what came to be known as "Yahoo-Yahoo" involved diaspora citizens being extorted out of their money under various guises and pretenses in Nigeria. Extortion permeated the system to the extent that even diaspora-building schemes became multibillion-naira fraud schemes. Those back home began to demand full entitlement to be served and taken care of, and if those in the diaspora faced financial challenges and couldn't assist on some occasions, it was met with resistance. As a result, those in the diaspora started to observe and analyze the approach and lack of transparency among those back home. It became clear to them that those back home would never appreciate anything less than their expectations.

The sacrifices made by those in the diaspora, working day and night to pay their bills and support their families, went unappreciated. This situation has led to serious conflicts between families, friends, and those in the diaspora. The Nigerian economy continues to decline, and the hardships faced by those in the diaspora are proliferating, with no signs of returning to the relatively better conditions of the early 1990s. Back home, the lack of transparency has become ingrained in the culture. As a result, those in the diaspora are no longer susceptible to falling prey to or accepting the culture of their homeland. Therefore, the only feasible and viable solution is straightforward. The diaspora community back home needs to understand, be open, be honest, empathize, and recognize the difficulties faced by individuals in the diaspora.

If so, perhaps trust between individuals in the diaspora and those back home can be restored, allowing support and cooperation to continue incrementally. The universal commonality among all these waves is the reason they are seeking a better life elsewhere, as the Nigerian government has consistently failed to meet the people's expectations. The majority have lost faith in their country's future, as progress tends to be reserved for those who belong to the elite. *As Aminu Kano famously said, "Nigeria will not know peace until the son of a nobody becomes somebody without knowing somebody."*

But what defines a "better life"? For many, it's the ability to live abroad and provide for family members back in Nigeria, often through financial support. What was once considered an act of kindness or compassion has become an increasingly expected behavior. Subsequently, this has led to what can be referred to as the "entitlement era," in which Nigerian families view remittances as an entitlement rather than a privilege.

This is not a new attitude among Nigerians; it is prevalent among diasporas sending money back home to relatives in various countries. And first-generation migrants came from small hometowns to work in cities and support their families.

BLESSINGS TO SHARE!

We, the Diasporans, each have a unique journey that has brought us to where we are today. Some traveled by road, others by air or sea, and some used a combination of these modes of transportation. For some others, the journey to the "promised land" was rapid and uneventful, but for others, it took decades and was perilous. Regardless of the time, the intention was always the same: to get to a place where one could live, work, and provide for their immediate and extended family. A few have arrived in this land of opportunity, but some are still on the journey, and some have perished. The physical, emotional, and psychological cost of making it here is humongous. But it speaks volumes of our resilience, patience, and determination as diaspora members.

Upon arrival, we had to navigate the process of obtaining residency and work visas through the immigration system. Regardless of the country, this is not a straightforward process. Some will take a few years, while others may take many years, as each case is unique and the requirements vary by nation. Some have successfully obtained their residence and work permits, while others are still waiting, and some have been denied, leaving them uncertain about their future.

I remember the first time I informed my cousin I couldn't wire him any money. He was quiet at the other end of the phone, a more serious silence than words. Not angry, not exactly, but disappointed. The slashing kind. I had lost my job, was behind on my rent, and could not keep up with my utility bills. None of that mattered. I was "abroad," and that alone was supposed to mean abundance.

I left home years ago with many responsibilities, including a wife and three sons to care for, but to him, all that mattered was that I was Dr. Eric, and I lived abroad and seemed to make a lot of money. That moment stayed with me not because of what was said but because of what it revealed. There was an unspoken belief that being in a foreign country automatically meant you were rich and that wealth was not yours alone to manage but to distribute.

Often, family members do not care how you are surviving life in another man's country; they never ask how you manage through life, and they often care about what you can send home. And it's not just my story. It's the story of the millions in the diaspora, men and women grappling with the weight of cultural expectations, guilt, love, and the complicated web of entitlement that spans oceans.

On the other side of the world, families back home bear their crosses: poverty, limited opportunities, and the hope that someone "out there" will be the breakthrough they so desperately need. In nearly every third-world household, there is a thread in common: someone has "made it" abroad. The instant you arrive in a foreign land, all bets are off. He is the anticipated messiah, unofficial patron, and ever-present cash machine.

Failure is no longer a personal issue; it is a communal, predictable, and often manipulated phenomenon. On the other hand, numerous people in the diaspora bear unseen burdens: the expectation of giving, the shame of saying no, the economic burden of unsustainable aid, and the sorrow of betrayal and exploitation. Bridging Expectations is a conversation that has been years in the making. It is about the tender struggles, the fragile connections, and the price tag of giving without restraint. And it is not that; it is about what it means to shamelessly, unapologetically pursue one another.

Mending, healing, and rebuilding with respect, boundaries, and integrity is what this is about. We can love our people and not lose ourselves in the process. We can give without being used. We can change the story.

People who Leave Home often face quiet sacrifices; they can't complain for fear of being deemed unkind or selfish. Some sacrifice their well-being by working 2 to 3 jobs to meet unrealistic expectations back home. When you leave the village, town, or country, the "black tax" is levied by friends and family.

I wrote this book for my brothers and sisters, hardworking immigrants who send money home monthly while skipping meals. It's for the family member back home who doesn't understand why the subsequent request isn't answered with a "yes." It's for all of us caught between love and obligation, generosity and boundaries, and hope and disappointment.

This book is not a condemnation but rather an invitation to learn, reflect, and reimagine what help can be in a healthier, more sustainable way. It requires difficult conversations, healing the sense of entitlement, and bridge-building —both monetary and emotional, encompassing empathy, empowerment, and mutual respect. ***SO, WHAT'S THE PURPOSE OF THIS BOOK?*** *The purpose of Bridging Expectations is threefold:*

1. **To bring awareness** to the deeply rooted entitlement culture that exists in many third-world societies and how it affects both those at home and those living abroad.
2. **To validate the lived experiences** of people in the diaspora who silently carry the emotional, financial, and psychological weight of unrealistic expectations.
3. **For presenting practical tools and solutions** for forging healthier relationships sustained not by compulsion or obligation but by shared respect, understanding, and empowerment. For bridging cultural, emotional, and relational divides between those residing "away" and those still "home." This book challenges old narratives and offers new, transformational perspectives that support growth, self-determination, and dignity on both ends.

I remember sitting on the floor of my apartment with dimmed lights, not for ambiance but because I was trying to cut corners on the electric bill. My phone buzzed; it was a WhatsApp message from home:

"Abeg bros, send me something small. Anything. You dey Australia."

Sadly, I had just spent my last $20 on groceries. I read the message and felt a sense of guilt. He was unaware of how hard things were for me and how I juggled two jobs just to keep up with the bills. It is a pity, you know.

-Emeka (Australia)

Chapter 1

WHAT YOU NEED TO KNOW ABOUT ENTITLEMENT

Before repairing frayed bridges, we must look closely at what broke them. One of the most unspoken yet deeply felt fractures is entitlement, an unseen shackle that binds love to expectation, sacrifice to bitterness, and generosity to resentment. Entitlement is not always evident. It doesn't always include loud voices or overt requests. Sometimes, it's the sigh you hear after the phone call.

The passive-aggressive "Is that all?" when you send money, the sudden withdrawal of affection when you say no. The quiet resentment festers beneath every family gathering, phone conversation, or unfulfilled request.

WHAT IS ENTITLEMENT?

Entitlement is a deeply ingrained belief that someone deserves your time, energy, or resources simply because of who they are or the nature of your relationship with them. Not because they've earned it, asked respectfully, or considered your situation, but because it is expected.

It often sounds like:

- "You're doing well over there. You should send something."
- "What's the point of being abroad if your people are suffering?"
- "Don't forget where you came from."
- "You make over $25 an hour; you should be able to buy the 50,000 naira Aso-ebi."

Effectively, entitlement murders empathy and replaces it with pressure. Entitlement makes achievement communal debt. It eradicates the reality of struggle and assumes abundance using geography. That is, it is not a question of whether you can give but that you must, at any cost. So, how does entitlement manifest in third-world countries? In most developing nations, a harsh economic reality prevails: unemployment is widespread, wages are low, and corruption, infrastructure deficiencies, or educational limitations generally limit growth opportunities.

These conditions give rise to a survival mentality; anyone "doing better" is a lifeline in survival mode. So, if one of your relatives goes abroad or to a wealthier city, they're instantly higher. And this increased status comes with expectations: school fees, hospital bills, rent, business capital, or even "something for the weekend." No one asks how you're coping, if you've paid your bills, or if you've eaten. The assumption is simple: you have, so you give.

In this environment, generosity becomes a performance. If you give too little, you will be judged. If you say no, you're condemned. If you delay, you're selfish. And if you stop giving altogether, you're branded an ingrate who has "forgotten home." In many developing countries, economic hardship is often passed down from one generation to the next. So when someone "makes it," whether to the city or overseas, they're seen as the ticket out. Suddenly, you are responsible for school fees, hospital bills, rent, funerals, weddings, and even business ventures you never agreed to fund.

There is little conversation about your capacity and even less empathy for your journey. If you say no, you're labeled proud, wicked, or "lost to Western ways." Entitlement is evident in family group chats filled with subtle guilt trips and social media posts that indirectly shame those abroad, including monthly expectations that go unspoken but are deeply felt, as well as manipulation disguised as emergencies.

This is not to say that every request is malicious, but the frequency, tone, and pressure often create a system of emotional taxation that is difficult to escape. Entitlement becomes normalized through unreasonable cultural expectations, as the firstborn, the abroad cousin, or the one who made it out of the village comes with its job description and responsibilities you never applied for.

Religious manipulation is another tool used frequently; scriptures like "Give, and it shall be given unto you" and, You are blessed to be a blessing" are used to guilt people into unsustainable generosity. The other is family coercion, under which group texts, last-minute emergencies, and public humiliation are typically employed to govern giving. Individuals are targeted within family gatherings, reinforcing that giving is your role.

THE HISTORICAL AND CULTURAL ROOTS OF ENTITLEMENT

Entitlement did not take place overnight. Communitarian values were at the center of life in most African communities. Inhabitants of villages lived on the principle of "if one eats, then we eat." It was an excellent, very humane value system based on interdependence.

But modern circumstances have tainted that ideal. It is no longer "We rise together" but "You rise, and we wait." What was once an overall sense of shared responsibility is now an expectation on the individual, with the individual caring for everybody.

Add in historical trauma, colonial disruption, failed governments, war, and poverty, and it's easy to see how hope was redirected from systems to individuals. When governments fail, families turn to their own. And those who leave are seen as the "escapees" tasked with funding everyone else's dreams. But here's the problem: People in the diaspora are not escapees. They are survivors. And often, they are barely surviving.

Let me share a short story about Tunde and the 10,000 Naira that cost everything. Tunde had just finished paying rent for his tiny shared apartment in London. His monthly transport card had drained most of what was left in his account.

The only food in his fridge was leftover rice, two eggs, and half a bottle of ketchup. But it was fine, he told himself. The month was almost over. He could survive a few more days. That's when the message came from his brother back home in Nigeria.

"Guy, things are rough here. Just send me something small." Tunde didn't think twice. He sent 10,000 naira, his last cash, hoping it would be of some help. But the response shocked him.

"Na, 10k you send? You dey whine me? What can this nonsense do? You wey dey chop pounds there. You no get shame?"

Tunde stared at the message. His eyes stung, but it wasn't just exhaustion. It was a betrayal. His brother didn't know or didn't care that that 10,000 was all he had left. He said that he had skipped meals to send it. Behind his UK postcode was a young man battling isolation, work stress, and the crushing weight of always being "the helper." He deleted the message. Sat in silence. Then, he cried not because he regretted giving but because he finally saw the truth: it was never enough.

EMBRACING YOUR CALLING!

Others are born with this habit of serving or giving, but are taught it through practice and learning. "From him who has been given much, much is expected." We are human beings, and our moral duty, regardless of our situation, is to serve people. It is not about having, but having the will to share. People would complain, "I don't have enough, so I can't share." Let us share regardless of our situation. If you have one, share one. If you have two, share two. If you have three, share them.

You don't need to have thousands or millions before you start giving. *If you only give when it's comfortable, you're practicing personal conviction.* But giving, even when difficult, reflects a more profound, divine conviction. Some people give with conditions or for self-gratification, but true giving should be unconditional. You may disagree, but I challenge you to look deeper into your soul. You'll find that *we don't deserve all the blessings we enjoy—it's by grace, not because we've earned it.*

Moral giving is possible with ethical intelligence. If giving is made everything about us, then its true purpose is forgotten: to honor the Maker of everything. In real life, unconditional giving ushers in higher blessings and true fulfillment.

Giving is not always in the form of funds; it can also involve giving your time, talent, or treasure. By giving your talent to bless others, you can experience the joy of what you were sent here for—your calling or your why. I pray that you will discover your calling as you accept it, for it is the source of peace that surpasses human experience and knowledge. It demands deep contemplation of life.

THE POWER OF TRUST AND HOW THE DIASPORANS GOT IT WRONG

Such distrust has a considerable impact, and putting it in terms of the amount lost due to it makes us realize how worse things are. If proper belief had existed, everything would have flourished, and the world would have become such an excellent place. However, we have not achieved this due to denial and the empathy rooted in our souls as people who value the principles of Ubuntu, which teaches, *"I am because we are."*

Let me clarify, the lack of trust is a cancer that infects and permeates our cultural landscape. This behavior that has been institutionalized is now the norm, facilitated by greed, evil, hatred, pride, prejudice, and selfishness. It has been built up over generations as this culture, each generation developing its definition of this behavior. Cultural change is complex and ingrained, making it challenging to reverse.

Corruption, a byproduct of this institutional activity, has become a way of life in most countries, including Nigeria, to the extent that it has become an integral part of the culture. Public denunciation of the same makes one a deviant, as one is viewed as gullible or "Mugu" (one who is considered foolish). Corruption in Nigeria has its corrupted language. Bribery, for example, is now more commonly referred to as "appreciation." The fact that language has changed reflects how deep-seated corruption is in the culture. It is something that the younger generations have grown up with and now expect—they believe every favor or assistance rendered to them must be rewarded.

This mindset has extended beyond Nigeria's borders, infecting even diplomatic circles. For instance, in some Nigerian embassies abroad, renewing a passport can entail "appreciating" the officials; otherwise, you will not receive your passport. If corruption could be enshrined as a right in the Nigerian constitution, our legislators would likely pass it into law.

I used to blame the government for this behavior, but now I see that citizens themselves play a significant role in perpetuating it. Here is an instance: if you need a product, such as cement, in Nigeria, and you ask a relative the price of cement, you'll receive a circumlocutory or extended explanation. If the cement price is N5,000, your relative may quote N6,200 per bag and pocket the difference.

Similarly, let's suppose that as a diasporan, you import merchandise, such as cement, to Nigeria for sale and ask your relatives in Nigeria to assist you in marketing the merchandise. Your relatives may downplay the market demand and price.

The feedback will indicate that no one buys cement anymore, creating an opportunity to capitalize on the market. They'll offer a lower price than the actual market value to make a profit. This culture extends to other transactions as well. For instance, if you want to sell chicken, you might be told that people no longer eat chicken, all to manipulate the situation for personal gain.

As diasporans, we are often compassionate and practical in helping family members back home. We provide them with resources and opportunities, only to find that they expect us to do everything for them. I call this the "do-it-yourself" mentality, where we in the diaspora are expected to do the heavy lifting. When we stop, our family members back home often complain angrily.

Part of the problem is that we're applying Western approaches to a situation that requires an Eastern cultural perspective. I advise stepping back and observing what's happening back home before reacting. Adopt a "submarine" approach—stay below the surface and only reveal what's necessary.

Take care of yourself before you put everything on the line for others. Who will bail you out if you encounter financial or health problems? Relatives likely expect you to bear their load, even when you are not faring well in the diaspora. The stress of getting relentless demands can lead to severe illnesses like stroke, breakdown, or depression. You are the only one *who* truly knows what you're going through, so take care of yourself. Trust is essential for us in the diaspora. It is the cornerstone for successful relationships and deals. We must establish trust among ourselves and our family back home to build genuine relationships founded on love, peace, and harmony. It is our responsibility to ensure that our relationships with those back home are rich and free from malice, hate, and bigotry.

However, until our relatives experience the realities we face, their mentality may not change. They see their situation as one of entitlement. The diaspora has long approached this with a Western mindset, which has led to frustration and disappointment. Be cautious about the information you share with family members, as they often read too much into everything.

My advice is to go silent for a while—let them reach out to you when they haven't heard from you in some time. The best solution is understanding that all dealings with relatives back home are often transactional. Accepting this reality may save you from frustration. It's unfortunate but true. When you don't give people space, you invite trouble. The less they know about your life, the better.

Your progress is often seen as being due to *luck or favoritism*, even when you work hard. This is the truth and the reality. Many relatives act in their self-interest, even at the expense of close family members. They play blame games to their advantage, and it doesn't matter if it's their son, daughter, mother, brother, or sister—they'll throw anyone under the bus if it benefits them. Diasporans should be aware of this common practice. To maintain peace of mind, avoid asking relatives back home for help, even with something as simple as time. Doing so can create unnecessary problems and heartache. You may feel indebted to them, even if you've paid for their services, because of their sense of entitlement.

MY TAKE

Entitlement is never so much about quantity as it is about attitude. The assumption. The belief that someone else's success belongs to you to take. But the truth is, you can't take from an empty well. You were never meant to be someone's hero. Love gives. But love respects. Support is okay. Support should never be silent servitude.

We must learn to unlearn that "no" is bad and to love the freedom that comes with healthy boundaries. Lastly, good support is not controlling but reciprocal. Unless we address entitlements directly, we will continue to build bridges on weak foundations.

Chapter 2

THE DIASPORA EXPERIENCE

From the outside looking in, life in the diaspora appears to be a dream: large houses, shiny cars, clean streets, and constant electricity. Social media paints it well: group photos in winter jackets, holiday shopping, family outings, or posts captioned "grind now, ball later." But behind those curated moments are layers most people never see: isolation, exhaustion, two or three jobs, student loans, unpaid bills, mental health battles, and the crushing pressure to "make it," not just for yourself but for everyone watching back home.

THE REALITY OF LIFE ABROAD

When someone leaves their country for "greener pastures," especially from a developing nation, they're often seen as having escaped. People rarely understand that the grass may be greener, but it comes with its thorns.

Here's what life in the diaspora looks like:

- Rent or mortgage that swallows half your paycheck.
- Working long hours, sometimes nights, weekends, and

holidays, to stay afloat.

- Balancing school and work to chase the dream.
- Cultural shock, racism, language barriers, and homesickness.
- Being "present" at home through money transfers while absent from your life.
- Many people abroad live paycheck to paycheck, not because they are irresponsible, but because surviving in a foreign land is financially, emotionally, and mentally expensive.
- Immigrants carry the pressure of two worlds: one that demands they prove themselves in a foreign land and another that waits back home, with hands stretched out, calling them "our helper," "odogwu, "our sponsor," or "our ticket out."

THE ILLUSION OF WEALTH AND SUCCESS

The illusion of plenty is perhaps the strongest stress generator for diaspora communities and people in their country of origin. You're in the U.S. You should be balling. You are in the UK. What does 50,000 naira mean to you? You're spending pounds. Not only are such statements false, but they are also offensive. They reduce entire careers of effort and sacrifice to exchange rates. They dismiss personal challenges with a wave of assumptions. They measure love in money sent and worth in remittances.

People often overlook the sacrifices made behind the scenes, such as skipping meals to afford a transfer, forgoing self-care to accommodate a family request, and missing birthdays, weddings, and funerals, not by choice, but because time off is unpaid and flights are unaffordable. Depression, loneliness, and burnout are silently endured because one is expected always to be strong.

Many in the diaspora live with invisible burdens. They're homesick but can't afford to visit, weary but can't afford to rest, and struggling but can't afford to stop. Through it all, they are still expected to send money, fix problems, and be the anchor for everyone else.

THE EMOTIONAL AND MENTAL TOLL OF ENTITLEMENT

Constant giving without replenishment can lead to financial and emotional bankruptcy. The mental strain of always being "the one who helps" is heavy. Many people abroad suffer in silence. They fear being judged as stingy or ungrateful. They overgive, overcompensate, and overextend.

Some people go into debt to maintain the illusion of success for their families. Picture yourself with no food in your refrigerator but still sending money because someone in the home told you they hadn't eaten. Imagine getting yelled at for not sending money, even though you've just been laid off. Imagine being informed that you are a terrible person for not attending a family reunion while your rent is past due.

That is the psychological cost of entitlement. It is what produces the dynamic in which people must prove their love in financial terms or risk emotional ostracism.

Over time, this leads to:

- **Resentment:** You catch yourself fearing home phone calls or messages.
- **Isolation:** You feel that no one gets you.
- **Anxiety:** You freak out whenever you see a transfer request.
- **Shame:** You feel like you're letting everyone down when you're doing the best you can.

THE LONELINESS OF BEING "THE ONE WHO MADE IT"

Being the successful one comes with a unique kind of loneliness. It's isolating. You're celebrated when you send money, but forgotten when you're struggling. People remember your blessings but not your burdens. You're never allowed to be tired and never allowed to be broke. You're never allowed to say no. You become a symbol, not a person. And symbols don't get to have boundaries.

But here's the truth: **You are not just a bank, you are not a savior, you are a human being.** And your worth is not measured by how much you give or how many people you support. *You have a right to rest. You have a right to boundaries.*

You have a right to live your own life, not one dictated by other people's expectations.

THE ILLUSION OF THE LONG HAUL AND WHEN THE YEARS DON'T EASE THE WEIGHT

It's often assumed that the longer you live abroad, the easier things get. Wealth follows the passage of time. Plenty is the harvest of experience. If a person has been overseas for over 10 years, there is an element of expectation that comes with that. For some individuals, however, such as Nkem, time only makes things weigh more. Nkem left Nigeria at age 24 when she was young, ambitious, and idealistic; she embarked on a journey to make her people proud. In her imagination, America was a place where streets were paved with gold and doors were wide open, full of possibilities for all. What she encountered was infinitely less than the promise.

Those early years were challenging, marked by part-time jobs, attending night school, and dealing with immigration issues. But she struggled through it. She finally obtained her papers, secured a good job, and purchased a small home. But the milestones were never just hers to celebrate. With every step forward came a flood of new responsibilities.

Nkem became the family's lifeline. She paid her siblings' university fees, sent monthly parental feeding money, sponsored cousins' visas, and sent money every month to her sister for upkeep, even when it meant going without. When her father fell ill, she covered the hospital bills. When her sister got married, she paid for the wedding. When her younger brother lost his job, she wired him money to "start something."

She had spent twenty years in America, yet she had no savings, retirement plan, partner, or children—just exhaustion, pressure, and silence. The loneliness crept in slowly until she broke down at work one day. She couldn't stop crying. Her manager sent her home. It was the first time she admitted it out loud: *she was burnt out.*

The therapist she finally visited asked her, "When was the last time you did something for yourself?" She couldn't answer. It had been so long. Nkem's story is not uncommon. Many in the diaspora, especially women, spend decades building others' lives while slowly losing their sense of identity. They live with the unspoken grief of missed milestones, lost dreams, and relationships that fade under the weight of responsibility. They send money but receive silence. They show up for others but have no one to lean on. And as the years go by, the question that haunts them is, *Who shows up for the giver?*

THE INFLUENCE OF THE DIASPORANS: FAMILY AND GOVERNMENT

Addressing the challenges at home requires a collaborative approach that involves family members in the diaspora and their relatives back home. Relying solely on government intervention is neither practical nor sufficient. Diasporans have already assumed the responsibility of providing essential services and infrastructure for their families, compensating for the inadequacies of governmental efforts.

Diasporans contribute by financing projects such as constructing new homes with reliable water sources (boreholes), reliable power sources using generators, security through security agents, transport provision, and sponsorship of quality education, health, agriculture, and infrastructure construction. These activities effectively meet the basic responsibilities that governments should undertake.

Nevertheless, the reality on the ground is very different and problematic. Among the main issues is project funds being embezzled by family members. In such situations, most family members often become victims of their greed, spending money on themselves rather than saving it for its intended purpose.

The feeling of entitlement also make things more difficult for other family members because it creates dishonest transactions and unfulfilled promises. Such a culture creates tension, distrust, and an enormous disconnect between diasporans and members of the family. Consequently, most diasporans have shifted their trust from relatives to friends or strangers to manage their projects.

While this approach might bring peace of mind, it is

unfortunate that this has become the situation. The opposition from family members, who fight against the wisdom of keeping outsiders at bay, only intensifies the deep-seated distrust. Despite that, diasporans are forced to swallow hard in acceptance to attain development and stability.

If the relatives fulfilled their duties and diasporans continued to assist, the government's role would no longer be needed. Such collective thought could compel the government to adopt new, more effective governance measures. It would signal that the people, both at home and abroad, are taking ownership of their destiny, thereby motivating the government to reform its approach.

Diasporans must continue to do their part, recognizing that the call for assistance extends across multiple sectors, including health, education, security, agriculture, commerce, trade, energy, and policy reform. The appeal to help the nation's resurgence is ethical and moral, and the moment to act is now.

Change needs to be participatory, not permissive. Diasporans must be prepared to roll up their sleeves and make the sacrifices necessary for progress. Decisive, direct actions that can seem as difficult as attempting to cram a square peg into a round hole, but are essential for advancing change.

The loss of opportunities due to the disconnect between diasporans and their native counterparts is a critical concern. The initial enthusiasm with which diasporans invest in their families, through consistent remittances or otherwise, has been curtailed due to chronic gaps and trust violations. Such a decline translates into forgoing immediate revenues to the recipients and an indirect economic impact on the nation. Loss of confidence leads to missed opportunities that would otherwise have solidified individual and national wealth.

It is essential to point out that the diasporans' silence shouldn't be interpreted as stupidity or cowardice. They are aware of the dynamics at play and fully understand the problems, even if they do not vocalize their concerns.

THE MISSING REALITY BETWEEN PEOPLE IN THE DIASPORA AND NATIVE HOMELAND

Diasporans supporting their families often face common challenges, including *financial strain.* Supporting family members in another country can be financially burdensome. Diasporans usually work long hours, take on multiple jobs to cover expenses, and send money back home. Another challenge is the cost of currency transfer and exchange fees.

International money transfer fees, exchange rates, and currency conversions are costly, reducing the funds diaspora members can send to their families. The majority come from countries with unstable economies; that means the money they send home fluctuates in value every time, affecting the purchasing power of their loved ones. The second is emotional suffering; physical separation from loved ones results in emotional stress. They may experience feelings of guilt, anxiety, and homesickness as they are unable to be present for important family events or provide immediate support during times of need. Another issue is dependence and expectation; families at home may become excessively dependent on financial support, which creates an expectation.

Individuals are subsequently forced to regularly fund their wishes and needs, putting pressure on them. There is also no appreciation. The givers are not typically appreciated, and their significant sacrifices usually go unseen by friends and families back home. Their work is not appreciated or given its due value, which can generate resentment and tension in the relationship.

What drives the tension between families and friends back home and those living in the diaspora?

1. *Financial Expectations:* People's expectations of receiving money from the diaspora at home have enhanced over the years. Families and friends who have grown accustomed to diaspora financing have often not made an effort to be more productive or independent, but had expected to receive money.

2. *Lack of Appreciation:* The sacrifices diasporans make, working and suffering overseas to care for their families

in their home country, often go unnoticed. This lack of appreciation can generate resentment and bitterness.

3. ***Transparency Issues:*** The lack of transparency in financial matters and project execution back home can breed mistrust and suspicion. Diasporans may question how their financial contributions are utilized and whether they have a positive impact.

4. ***Cultural Differences:*** Since they have been exposed to other cultures, there may be cultural differences and expectations between the diasporans and individuals at home, which can lead to tensions and misunderstandings. The diasporan' would want certain behaviors and values to be maintained, but people at home would feel differently or prioritize how things should be done.

5. ***Unrealistic Expectations:*** With some back home increasingly perceiving entitlement, there may be unrealistic expectations for financial support or goods from the diasporans. These can pressure relations and lead to tensions if they are not met.

6. ***Absence of Empathy:*** There is a lack of empathy, and no attempt is made to understand the problems faced by diasporas, such as cultural or economic challenges, that contribute to conflict. Empathy does not facilitate gap closure or the seeking of commonalities.

7. ***Economic Challenges:*** An underperforming economy

combined with decreasing opportunities for labor and economic mobility can depress and infuriate citizens in the home country. This is perhaps due to their association with diasporans, who are presumed to have greater opportunities in foreign countries.

Recognizing the sacrifices of diasporans is crucial, as it deeply affects their relationships with loved ones back home.

Here are some ways relationships are affected:

1. ***Frustration and Resentment:*** Frustration and resentment are experienced whenever diasporans feel that their hard work and sacrifices are not appreciated or acknowledged. They feel frustrated thinking about why they work so hard but are not appreciated.

2. ***Emotional Disconnection:*** Ungratefulness leads to estrangement among diasporans and their relatives and friends. They become withdrawn and are no longer interested in maintaining contact because they feel they are not appreciated for their input.

3. ***Strained Communication:*** Disappreciation can create strained relationships between the diasporans and their friends and relatives back home. Disappreciation may lead this group of people to believe they can no longer freely express their feelings and thoughts, which can lead to

conflict and misconceptions.

4. ***Lower Incentives:*** If diasporans do not feel appreciated, their incentive to continue supporting their families and friends in their home country will decrease. They will question why they are making such significant sacrifices and be less likely to provide either economic or non-economic support.

5. ***Poor Sense of Belongingness:*** The diaspora's failure to appreciate its contributions hinders its development of a sense of belonging. It tends to feel disconnected from its place of birth and is often unable to forge strong connections with its parents and friends who remain in the country of origin.

6. ***Destroying Trust and Support:*** Ingratitude results in others losing support and trust in relationships. Diasporans are hesitant to offer support or engage in exchange in the true sense because they feel that their contributions are not valued.

The sacrifices of their diaspora counterparts need to be acknowledged and appreciated by their families and relatives in their homeland. An expression of gratitude, acknowledgment of their service, and frequent, sympathetic interaction can do wonders in consolidating relationships and bridging the gap caused by feelings of abandonment.

Those back home must value and acknowledge the diaspora's sacrifices for many important reasons.

1. *Emotional Support:* If their sacrifices are acknowledged and appreciated, it confirms their endeavors, as their hard work is recognized and valued, thereby increasing their morale and overall well-being. Appreciation gives emotional support to diaspora communities. It legitimates their sacrifices and confirms their experiences, making them feel emotionally stronger and resilient.

2. *Deeper Relationships:* Gratitude creates deeper, more lasting relationships. If the diaspora population is valued for their sacrifices by their compatriot friends and families back in the country of origin, the bond is deepened. It can be converted into better two-way communication, greater understanding, and greater feelings of togetherness.

3. *Constant Incentives and Support:* Appreciation is an incentive. Diasporans are motivated to support their peers and close relations in their country of origin if they feel appreciated. It strengthens their determination and allegiance towards their nearest relations.

4. *Less Frustration, Misunderstanding, and Resentment:* Being unappreciated breeds resentment and frustration. Appreciation of what diasporans do, prevents these feelings from happening. It creates an atmosphere where both feel

appreciated and respect one another. Not acknowledging and appreciating other people's sacrifices in the diaspora can result in misunderstandings. Acknowledgment and appreciation help prevent such negative feelings and foster a more friendly and cooperative relationship.

5. ***Sense of Belongingness:*** Gratitude fosters a stronger sense of belonging. Appreciation for the sacrifices made by those abroad on the part of their people at home strengthens their place in society and within their families. It allows them to remain firmly attached to their cultural identity and origins.

6. ***Mutual Respect and Empathy:*** Appreciation is followed by empathy and mutual respect. Suppose the diasporans are appreciated for their sacrifices in their countries of origin, their compassion for the hardships of the latter, and their contributions to the labor force. This creates admiration for their lives and gives rise to support and empathy for them. It also creates mutual understandingIt closes the experience-perception gap by enhancing the bond of mutual understanding, enabling empathy.

Appreciating and acknowledging the sacrifices of diaspora workers, their families, and their local support groups can strengthen relationships, provide emotional support, and foster an atmosphere that fosters ongoing support and empathy. This two-way system effectively closes the gap and is suitable for establishing genuine relationships, regardless of distance.

Appreciating the diaspora's sacrifices brings many benefits to both the country and their loved ones back home for many reasons:

1. ***Stronger Personal Links:*** Appreciation and recognition strengthen the emotional bond between diasporans and their closest relatives and friends in their homelands. It makes them feel closer and results in close, tight relationships.

2. ***Inspiration and Motivation:*** Diaspora individuals are motivated to fight and sacrifice for their dependents when appreciated and acknowledged. They are even encouraged to persevere without losing hope because they are certain their struggles are valorized.

3. ***Greater Support:*** Appreciation for sacrifices made in the diaspora encourages others to reciprocate support in any form possible. It is a two-way street, where both parties are willing to lend a helping hand in times of need.

4. ***Preservation of Culture:*** Appreciation for the sacrifices of the diasporans highlights the importance of cultural preservation. It deals with their contribution towards their culture and heritage, ensuring the continuation of customs and traditions. Generally speaking, recognition and gratitude of diaspora sacrifices for their fellow citizens in their countries of origin strengthen bonds, provide motivation and encouragement, engender empathy, and create a sense

of greater belongingness. It is enriching in both directions, promoting meaningful relationships and the welfare of both sides.

MY TAKE

Overseas life is not a fairy tale. It's an everyday struggle marked by sacrifices, difficult choices, and silent battles. It's time we start telling the whole truth. Not just for us but for the next generation. Because if we don't raise our voices, we will continue to bear burdens that were never ours to bear alone.

May this chapter be a mirror of the unseen, the undeserving, and the overworked. May it be a window to the people back home so that they can see that the people they admire overseas are also like them: striving, struggling, hoping, and doing their best.

Chapter 3

THE BURDEN OF GIVING

Giving is beautiful. Even the good book implores us to give and to carry each other's burdens. Giving is one of the most profound ways we show love, empathy, and connection. But what happens when giving turns into an expectation? When does generosity become your identity? When does saying no feel like a betrayal? Many of us become addicted to giving, fueled by the fleeting affirmation of being seen as saviors.

There is a thin line, though, for someone like me who was born with a life's calling to give shelter and sustenance to orphaned children, giving is part of my DNA. Unfortunately, entitlement makes giving a burden. To most of us in the diaspora, it has been a burden that they have borne silently, in pain and solitude, for decades.

THE CYCLE OF CONSTANT GIVING

It usually starts with love, a one-time gift, a transfer to help with rent, school fees, or a hospital bill. But over time, the requests become regular, the gratitude lessens, and what started as generosity morphs into obligation. Next, you become Father

Christmas, the genie in the bottle that must grant wishes as soon as the phone rings. An acquaintance recently told me how she spent 28 years sending items home. For the first time last year, she went home and didn't have a suitcase filled with goodies for her family; her father insulted her.

Being a giver is excellent, but the moment you stop or say no, you become the villain, the failure of the family. The obligation of being a giver makes you the provider. The one who "always sends something." The one who must never say no. The one who can never have problems of their own.

And so the cycle begins:

1. You give once.
2. They come back again.
3. You hesitate.
4. You're guilted.
5. You give again—to avoid being seen as selfish.

Eventually, you find yourself budgeting not just for your bills but for theirs, living not just your life but also carrying theirs as well. You start asking yourself, "When did I become responsible for everyone else's survival?" And in most cases, some of the takers do not appreciate your efforts; you become the butt of the joke as your hard-earned money is being spent at a pepper soup joint while you have to eat your last ramen noodle.

You are the hot topic at the next family gathering, and

your siblings are comparing you to Gregory, who moved to Canada and built a mansion for his parents in less than a year. Your generosity is overlooked, and your efforts are often met with silence or judgment.

STORIES OF EXPLOITATION, DECEIT, AND MANIPULATION

In some cases, giving turns into a tool of manipulation. When I conceived the idea for this book, I decided to gather stories from individuals in the diaspora and those who had returned "home," and the stories were the same. And these stories cut across cultures.

There are stories of people being misled about emergencies to obtain money. Of relatives faking sickness. Of siblings doubling the price of school fees. Of friends asking for business capital but spending it on luxury. Of family members building houses and never acknowledging the one who funded it, or misappropriating the funds for a business or building project.

A young man once sent money home to build a house for his retirement. When he finally returned after 15 years, he found an empty plot of land. His brother had squandered every dollar but smiled as he welcomed him back, saying, "You know how things are." Thankfully, his life was spared; there are cases of individuals who were not so lucky; they lost their lives after the thief was found, so they plotted to kill the "giver." This is the trauma of trust broken by entitlement.

HOW GUILT SUSTAINS THE ENTITLEMENT MINDSET

Guilt is the glue that holds this system together. For many in the diaspora, saying no feels like a betrayal. **The guilt is layered:**

- **Survivor's guilt** – because you "escaped" while others did not.
- **Religious guilt** – because scripture teaches you to give, bless, and help.
- **Cultural guilt** – because "we do not forget where we came from."
- **Family guilt** – because your success is seen as a family investment.

And so you give even when it hurts, even when you do not have. Even when you know the money will not be used wisely. Because to say no is to risk losing your place in the family, being shamed, insulted, and ostracized. Or worse, being labeled proud, changed, or heartless. But what is the cost of giving when it's rooted in guilt?

It's the slow erosion of joy. It's waking up each day with the weight of expectation pressing on your chest. It's looking at your phone with dread because every message might be another ask, another emergency, another obligation.

Giving from guilt is like pouring water into a bottomless pit; no matter how much you give, it's never enough. Eventually, you stop sharing with your heart and start giving with your hands shaking, your chest tightening, and your spirit whispering, "This is not sustainable."

Guilt-based giving breeds resentment. You begin to question your goodness. You start measuring love in transactions. You wonder if anyone sees you beyond what you provide. And that's when the giving turns dark, not because you don't want to help, but because you no longer feel seen. The cost of guilt is that it silences your truth. It keeps you in relationships that take without refilling. It traps you in cycles of emotional obligation. It teaches others that access to your heart means access to your wallet.

True generosity flows from love. Guilt mimics love, but it's counterfeit. It gives to be accepted. It helps avoid conflict. It offers to earn approval. But love—real love—gives from freedom, not fear. ***So the real question becomes:*** What would your giving look like if it came from a place of choice, not coercion? What would it feel like to give from joy, not pressure? Because the cost of guilt is too high. And your peace is far too precious to keep spending on expectations you were never meant to carry.

WHEN GIVING BECOMES EMOTIONAL BONDAGE

There is a fine line between generosity and emotional captivity. Giving should be life-giving, not life-draining. However, it becomes a bondage when tied to validation, approval, and the fear of rejection. And the most dangerous part?

People start to measure your worth by how much you give, not by who you are. Not by your presence. Not by your sacrifices. But by the last money transfer. Relationships become transactional, affection becomes conditional, and love starts to feel like a bill you must constantly pay.

Patel moved to Canada from Gujarat, India, in his early twenties. Like many immigrants, he arrived with little more than ambition and a dream. He worked long hours, first in construction, then as a delivery driver, while studying part-time to become a certified technician. Eventually, after nearly a decade of sacrifices, Patel secured a stable job, bought a small condo, and even sponsored his younger brother to study in Toronto.

From the outside, Patel's story appeared to be a success. But underneath the surface, the pressure was relentless. Every month, he sent money home to support his aging parents, his sister's school fees, and a cousin's wedding expenses. No one ever asked if he could afford it. It was expected. He was now "Patel Canada," the family's pride and primary provider.

The calls never stopped:

Dr. Eric Edokpaigbe Idehen

"Uncle needs help with his medical bills."

"Can you send money for the roof repair?"

"A small loan for your cousin's business, we'll pay back soon."

But the loan was never repaid. The expenses continued to grow, and Patel continued to give even when it meant skipping vacations, delaying his wedding plans, and putting off investing in his future. Then came the moment of collapse. One night, Patel sat alone in his car outside his workplace. His phone buzzed with another urgent and demanding request from home. That's when it struck him: he had not been inside his own house in five years, had not missed a day of work in months, and had not told anyone that he was depressed. He felt invisible.His value had been reduced to how much he could send, not who he was. Nobody ever asked, "Are you okay?" They only asked, "Did the money come through?"

ENTITLEMENT IS A GLOBAL PANDEMIC

There is a global sense of responsibility that exists beyond one culture, one nation, or one continent. Whether in Africa, Asia, Latin America, or Eastern Europe, the weight of being "the one who made it" is a global phenomenon. The demands will be of various kinds, but the same script: guilt, responsibility, and emotional exhaustion. Patel's account reminds us that behind every Western Union receipt is a tale of sacrifice.

Behind every monthly remittance is someone sacrificing others before themselves, again and again. So, how do you take back generosity?

Your goal is not to quit giving. Your goal is to give with purpose, boundaries, and clarity, transitioning from the family ATM to a role model of wisdom—a sustainable and emotionally healthy one. Giving does not have to be at your expense.

SO, WHAT IS THE REAL PROBLEM?

The average salary for a civil servant in their home country is a mere $30 per month. This inadequate income is earned through immense dedication and hard work, often accompanied by the frustration of enduring months of salary delays.

However, when individuals in the diaspora send remittances back home, usually five times or more than the average salary, their generosity is not always appreciated. Sometimes, their people back home want the diaspora to apologize for failing to fulfill their lavish expectations.

Their people back home have become so dependent on the diaspora's financial support that they have become complacent. They have not attempted to be creative and come up with entrepreneurial solutions and ideas for development. Such inaction discourages the diaspora from investing more in the homeland. The most annoying part is that, despite some people's financial potential back home, they act as if they are poor and still want the diaspora to send them money. This is exploitation since it profits at the expense of other individuals' goodwill in the diaspora. All these matters, therefore, must be addressed, and an attitude change must be fostered among the country's citizens.

Instead of relying solely on economic assistance, they must be self-sufficient and generate their employment. By promoting an entrepreneurial culture and innovation, they can promise growth and draw additional diaspora investment. Additionally, there is a need to promote openness and accountability in finance.

If individuals back home are financially capable, they should be honest about their resources and refrain from exploiting the diaspora for monetary gain. This will help foster trust and encourage more significant support from those living abroad.

WHERE THE DIASPORANS WENT WRONG

The effects of not being truthful about the struggle to make money outside are already experienced by diaspora individuals. It is a trend among those outside competing with one another in displaying their wealth and prosperity. Sadly, it has given the wrong impression that making money outside is easy for those back home. Despite reality, the illusion that it is laid-back and simply earning money outside continues. It is discovered in research that 8 out of 10 members in the diaspora suffer from relationship breakdowns as a result of numerous reasons, such as no trust, openness, or appreciation towards each other, and instead profound hatred and anger towards each other. The first key reality to accept is the challenge of earning money in different countries.

Unlike what others in the diaspora had described it to be, it is not such an easy thing to succeed in a foreign land. There are just too many issues that come in between, whether it is cultural and linguistic differences, limited opportunities, or bias. Immigrants are not always presented with readily available jobs, particularly in highly competitive fields.

It takes a phenomenally high commitment, work, and resolve to succeed in his field in a foreign land. Contrary to what others in the diaspora have described, succeeding in a foreign country is not an easy task. Most things come with obstacles attached to them, be it cultural or linguistic differences, restricted opportunities, or discrimination.

Immigrants are often underemployed, particularly in highly competitive and demanding occupations. It requires a high level of commitment, hard work, and resilience to excel in one's field in another country. Furthermore, the competition between diaspora individuals for what they have is the source of the misconception that easy money is readily available to them. Social media are breeding grounds for displays of prosperity as people showcase their lavish lifestyles and possessions online. In reality, it should be observed that such displays are not the true reflection of the reality of people in the diaspora, as they might be economically struggling, in debt, or otherwise engaged in terms of their finances, irrespective of the ostensible demonstration of prosperity.

A lack of integrity and transparency regarding the experiences of people in the diaspora is a breeding ground for distrust and strained relations. Suppose people at home feel that their diaspora counterparts are leading lavish lives at their expense without appreciating the sacrifices they make while experiencing hardships. In that case, resentment and jealousy may be caused.

This may lead to strained relations as people at home feel exploited or abandoned by their diaspora counterparts. Aside from the deep hatred and anger, the resulting feelings stemming from this reality underscore the negative impact of the shortfall in transparency and honesty. These deep feelings are rooted in disappointment or betrayal, as people in the diaspora are perceived as hiding the reality of their economic struggles and actual circumstances in their countries of origin.

Such feelings may result in deep relational ruptures, resulting in an unbridgeable distance between close acquaintances or within the family. Such issues require initiating dialogue and a culture of trust, transparency, and empathy between people in the diaspora and their people at home. Frankness in discussing the facts on the ground, particularly in connection with life in foreign lands, the challenges faced, and the sacrifices made, supports bridging the gap in how they are perceived. Diaspora people should be encouraged to share their experiences, achievements, and challenges.

Finally, diasporans should be grateful for any remittances they receive from their families and communities in the homeland. Expressing gratitude for financial support can help reinforce trusting relationships and prevent feelings of jealousy or resentment. Gratitude does not necessarily require financial support in the form of remittances; it also needs emotional support and a connection with loved ones.

Public campaigns also help break myths and inculcate an understanding. They can provide accurate information on diaspora issues and help overcome the perception-reality gap. Such campaigns may also ensure the free exchange of ideas and facilitate communications among the diaspora people and their kin in the country of origin.

In conclusion, the consequences of dishonesty regarding diaspora struggles to make ends meet are vast, leading to shattered relationships, distrust, and deep-seated hatred and resentment. The struggles of people in the diaspora must be embraced, honesty, transparency, and thankfulness must be cultivated in their lives, and via cultivation of openness in relations, learning, and consciousness, gaps between reality and perception can be bridged, healthy relations, as well as awareness between the diaspora communities and homelands, can be cultivated.

MY TAKE

You are allowed to give. You are also allowed to stop. You are allowed to care. You are also allowed to rest. True love gives, but it also respects. True family supports, but it also understands. True generosity flows from choice, not pressure.

As you read this chapter, ask yourself the following:

- Is my giving still rooted in love, or has it become a matter of survival?
- Have I given people the impression that I don't have my own needs?
- Have I set clear boundaries with those I support?
- What does healthy generosity look like for me moving forward?

Let this be the season when you free yourself from the burden of giving without boundaries. When you give from a place of wholeness, your support becomes empowering rather than enabling. And that's how we begin to bridge expectations truly.

Chapter 4

HOW DOES ENTITLEMENT AFFECT RELATIONSHIPS?

Entitlement exhausts emotional vitality, not just financial assets. It causes undue stress on relationships, turns them into responsibilities, and gradually erodes trust and love. If gifts are not reciprocated, expectations remain unchecked, love becomes conditional, and relationships will eventually deteriorate. Over time, feelings between family members become weak, and friendships strained, romantic connections collapse— all because expectations outweigh empathy, and entitlement replaces understanding.

Family is supposed to be a place of safety, unconditional love, and shared responsibility. But when one person becomes the provider and everyone else becomes expectant, love turns transactional. Many in the diaspora report feeling like outsiders in their own families. They are cherished when they send, forgotten when they do not.

They are appreciated for what they send and not for

who they are. When they return home, they are greeted with open arms, not because they are at home but because of what they might have brought. Even milestones come at a price. Weddings are now an expensive essential necessity. Funerals are an expense. Graduation of a brother or sister is an expense with no proviso. And if you dare to say "I can't" or "Not this time," silence follows, or worse, slander.

You are added to a WhatsApp group without permission to contribute openly, and can't leave. You are expected to give the most because God forbid you give the same 20,000 naira Ehi the teacher gives.

CHIAMAKA THE SILENT SISTER

Ada had always been close to her younger sister, Chiamaka. They shared everything growing up: clothes, secrets, and dreams. When Ada moved to the U.S. for school, she promised to always be there for her. For the first few years, she sent money regularly. Ada paid for Chiamaka's university fees, rent, and laptop.

However, as Ada's bills accumulated with student loans, car repairs, and rent, she asked Chiamaka to be more responsible. To get a part-time job, to be able to contribute already. Chiamaka ceased to answer messages. She would not answer calls, not even on Ada's birthday, she didn't send one single text message. Weeks turned into months. Months later, finally reached out to her, "You've changed since you left for abroad; you look down on us now." There is not a single word about the support year in and year out, nor is there any acknowledgment of Ada's struggles; instead, there is only disappointment that the money had slowed. Ada wept, not for the money lost, but for the sisterhood that entitlement had stolen.

Beyond entitlement in families, friendships can also turn transactional. Friendships are two-way transactions of laughter, memory, and support. Bring in the money factor, especially if the person is abroad, then goodbye, old pal, till he needs help. Hello, greetings are made, and demands for money are made at the end. You are a superhero if you help; otherwise, you are out of favor, unfriended, or blocked.

WHY TUNDE GOT UNFRIENDED

Tunde had just one childhood friend, who was Kwame. They had been close in secondary school, and even after Tunde moved to Germany, their relationship remained close. They would exchange voice notes jokingly, talk about the past, and fantasize about doing something together in the future. And then things changed for Kwame.

He lost his job, and Tunde sent money to help twice. The third time, Tunde explained he couldn't. He had just started a new job and was behind on his rent. The response? A long, passive-aggressive message.

"People like you forget about us after they start to gain financial success. You have let Western culture change you. You are not the friend I used to know."

Then there was silence. Kwame blocked him on WhatsApp. Years of friendship came to an end due to unrealistic expectations. He could not believe their relationship was based purely on money matters.

ROMANTIC RELATIONSHIPS AND EMOTIONAL WITHDRAWAL

Romantic relationships suffer, too. Partners back home often become emotionally distant once financial support is withdrawn. At other times, love is weaponized and kept alive only through the promise of monetary benefit. Some in the diaspora fall into the trap of being loved not for who they are but for what they can offer. And when they stop giving, the affection fades.

Other times, diaspora partners fear dating or marriage altogether, traumatized by past experiences where relationships were built on financial dependency, not real connection. Women are all lined up during the holidays to meet their ticket to Obodo Oyinbo (The White Man's House). Men end up with women who use them solely for their pockets and access to the blue passport.

There are stories of men and women who dated or

married a diasporan because of what they could get. What happens when the people who are supposed to be your safety net become the source of your deepest emotional exhaustion? That's what happens when relationships are built or rebuilt on entitlement rather than trust and mutual care.

The most painful part? Many in the diaspora can't even discuss these issues without being dismissed. If they express frustration, they are perceived as being "arrogant," "stingy," or "ungrateful." If they ask for sympathy, they are informed, "You made your choice to go abroad," or "You have no idea what's happening here."

Their emotional distance widens the gap between the diaspora and their people. Increasingly, they no longer share their problems. They censor themselves and become emotionally distant because relationships are no longer safe, not because they don't care.

THE STORY OF THE COUSIN WHO VANISHED AFTER THE WEDDING

Chidera adored her cousin, Onyeka. She practically raised him, paying his school fees with her income as a nurse in the UK, and consistently sent money for his upkeep. When Onyeka announced his wedding, he asked for help with the bride price, the hall, and the catering—Chidera paid it all. She even sent an extra 500 euros as a gift.

But she didn't get an invitation to the wedding. She only saw pictures online. When she reached out, Onyeka said, "You know you were too far away to attend, but thanks for all your help." That was the last time they spoke to each other. No thank-you card. No update. Just silence. Chidera struggled to realize that her support had been financial, not relational. She was no longer seen as family, but rather as a facilitator of their needs. To this day, there has been no closure. She has no idea what went wrong and how the good she did turned bad, destroying their relationship. Many others have shared a similar experience, including being ghosted after helping someone from back home.

I find it difficult to understand this behavior, but I can only assume that those relationships were built on selfish interests. Once they get what they want, they disappear. Selfish interests drive them, and once they get what they want, they vanish.

THE STORY OF THE UNCLE WHO KEPT TABS

Bashir had an uncle in Nigeria who would call him once every other week with some new money problem. One week, his car would be in the shop; the next, he had "obligations" for his neighbor's funeral expenses. Bashir finally started saying no to his uncle. He needed to prioritize his needs and save for his wedding. His uncle once sent his extended family an emergency message via group text, in which he wrote:

Whenever people are paid peanuts, they forget how

someone fed them. Bashir suffered deep sorrow, not just because of the allegation but because it undermined years of devotion, charity, and affection. Once the giving stopped, so did his devotion.So, how do you identify and heal transactional relationships?

Here are a few signs a relationship may be leaning into transactional territory:

- You dread hearing from them because you know a request is coming.
- You're only contacted when money is needed.
- Your boundaries are ignored or disrespected.
- You're made to feel guilty for not doing more.
- Your sacrifices are never acknowledged—only your current "lack."

If any of these resonate, it's time to pause and reevaluate the relationship, not from a place of anger but from a place of truth. These are some lessons I have learned over the years on rebuilding relationships with boundaries and honesty, and healing strained relationships due to entitlement.

1.	**Have the Hard Conversations**: Sometimes, people don't know how much you're going through until you tell them. Share your struggles, be honest about your limitations, and let them see beyond the currency exchange rate into your reality.

2.   **Define What Support Looks Like for You**: Support isn't always money. It can include mentorship, emotional support, assistance with job searches, or guidance. Let your people know that your love isn't limited to financial assistance.

3.   **Stop Over-Explaining Your "No"**: Your boundaries are not up for debate. You don't have to explain every financial decision you make. **Practice saying:**

- *"I'm not in a position to give right now."*
- *"I love you, but I can't support that at this time."*

4.   **Concentrate on Mutual Investment**: Ask yourself: Is this relationship in any way reciprocal with me? Not that it has to be financially, but is there respect? Support? Reciprocity? If not, it may be time to reassess the effort you're putting in.

5.   **Forgive, but Don't Forget**: Some relationships will never return to their former state. That's okay. You can forgive individuals without reopening the door. Healing also means safeguarding your peace.

Relationships are sacred, but entitlement corrupts them. It turns something beautiful into a contract, where love must be earned, and presence must be paid for. But you are more than what you send and more than what you can provide; any relationship that values your giving more than who you are isn't real. The relationship is founded on dependency and bound by obligation.

You deserve genuine relationships rooted in empathy, mutuality, truth, and freedom.

MY TAKE

Entitlement costs more than money. It warps the meaning of love and redefines relationship roles, from sister to sponsor, friend to financier, and partner to provider. Healthy relationships are built on communication, empathy, and support—none of which include silent expectations, love guilt, or strings of giving.

As you ponder, ask yourself. Which relationships of yours are transactional or one-sided? Where have I allowed guilt to dictate boundaries? Reframing love and attachment in terms other than economics. Be honest about your relationships and rebuild them. Define what giving is and what love is. You don't have to stop giving, you just have to give with wisdom.

Chapter 5

THE CITY SYNDROME
(Village to City Entitlement)

Entitlement doesn't only exist between continents. Sometimes, it travels much shorter distances, like the stretch between a rural village and a bustling city. Moving to the city is often viewed as a significant success in many parts of the world, particularly in Africa and South Asia. You don't need to leave the country to be labeled "the one who made it."

This is the Village to City Syndrome: *a dynamic in which people in the village look to their relatives in the city as their personal safety net, solution center, and emergency fund. The same pressure exists between the diaspora and their home country, as well as between urban dwellers and their rural roots; it is just closer, more frequent, and often even more culturally demanding.*

City life doesn't always glitter, but it shines to someone looking in from the village. It glimmers with hope, possibility, and the pretense of convenience. That illusion, however, creates an insidious pressure cooker of expectation that begins to dictate commercial relationships and identity.

That youngster, brother, or cousin who came to the city is no longer just their name. It turns them into an icon—a human wallet, a life ring for those at home. As time passes, the symbolism turns into an implicit contract. No words need to be spoken. It's just understood; you've left, so you owe. Whether or not you're thriving in the city doesn't matter. What matters is that you are no longer "here," meaning you're doing better. In this culture, geography becomes a form of currency.

The burden begins with small things, a call from your mother asking for help buying kerosene. Your uncle needs a bag of cement. Your younger sister is starting school, and the school fee is higher than expected. And then it spirals out of control. A new roof for the house. A loan for your cousin's business venture. A wedding gift for your friend. A generator for the local pastor.

Suddenly, you're not just helping out family, but you're helping out the whole village as well. You are the default option. The emergency fund. The solution center. However, with each act of giving, the line between generosity and expectation becomes increasingly blurred. What was once gratitude turns into quiet resentment if the support doesn't come on time or if it falls short of their perceived need.

The more you give, the more you're expected to give. And when you slow down, they say you've changed, the city has corrupted you, and you've forgotten your roots. This kind of entitlement is deeply cultural and emotional, and often reinforced by feelings of guilt. You're reminded of your parents' sacrifices to send you to school, the elders who prayed for you, and the neighbor who watched you grow up. You're told it's your turn to give back, but you don't realize that "giving back" has become a bottomless pit. It's even more painful when the ones demanding the most are the ones who never showed up when you needed them.

They didn't call to encourage you when you were jobless. They didn't check in when you were sick. They didn't ask how you were managing when your rent was overdue. But now that they believe you have "made it," their presence is loud, their messages persistent, and their needs endless.

There is such fine manipulation in this dynamic. If you do not help, you will be reminded of how others may perceive you as arrogant. If you hesitate, you are reminded of how they fed you or helped your mother. It is a moral obligation you never asked for but are now expected to repay. And if you stand up for yourself, you are selfish or unthankful.

This kind of emotional manipulation is hard to confront because it hides behind the language of tradition. "It's our way," they say. "We don't forget where we come from." "You are the hope of this family." And so you give, not because you can, but because saying no would mean dishonoring those words. You start shrinking your own life to support others. You delay your dreams. You say no to your needs so that you can say yes to theirs. You do it silently, sometimes bitterly, and always at a cost. You delay purchasing the new stove for your home because of the obligation; you shop at discount stores or only on sale racks to accommodate your needs.

Many people living in cities send most of their income back to their villages, never taking vacations, never investing in themselves, and often living in poorer conditions than those they support. They live in small rented apartments while funding house projects in places they may never return to. They avoid relationships because they fear being loved for their position, not for who they are.

They stop sharing their lives with family because every phone call ends in a request for something. What makes it even more heartbreaking is the isolation that comes with it. People in the city surround you, but you feel far away from those you call family.

You start building emotional walls to protect your heart. You start dreading home visits because of what will be expected of you. And if you do go home, you're no longer "one of us," but are now a stranger. Speak in different ways. Dress in various ways. And above all, we are expected to give in other ways. And yet, you love them. And you still want to give. But you do want to breathe, too. You want to give in freedom, not in fear. You want to support, not be stifled.

However, the middle path is impossible to tread. Because the problem is not given at all. Generosity is an admirable act. The problem is the imbalance. In the silence of your own needs. In the entitlement masquerading as culture. In the relationships that measure your love in dollars, not in people. But finding that balance feels impossible. The truth is, giving isn't the problem.

Generosity is lovely. The problem is in the imbalance. In the silence about your own needs. In the entitlement, masquerading as cultural principles. In the relationships that measure your love in terms of currency rather than connection. It takes courage to rewrite this dynamic. Courage to be honest, to say no without apology, and to reframe what helping entails.

It may be giving advice rather than dollars. It may be helping someone get employment rather than funding another loan. It may be investing in someone's growth rather than covering their deficit. It also means modeling a different way of being; boundaries are not rejection but respect for yourself and those you love.

People may not initially understand it. Some may even pull away. But those who like you for who you are, not your wallet, will ultimately be there. And those are the ones you want to nurture. The city syndrome doesn't have to define your story. You can live in the city and still honor your roots. You can rise without becoming everyone's ladder. You can love your people without losing yourself.

Giving should not feel like punishment, and family should not feel like a burden. It should feel like home.

THE WEIGHT OF "MAKING IT" IN THE CITY

City life presents its own set of challenges, including rising rent, lengthy commute times, inflation in meal costs, and labor market uncertainty. But those in the village don't see it that way. From their point of view, your place in the world shouts out luxury.

Do you live in Lagos, Accra, Nairobi, or Mumbai? They believe you are eating your fill, earning dollars, and driving a car, even if you are sleeping in one room with three others, just trying to get by. That assumption becomes an expectation. And before long, it becomes an obligation. Soon, each call from the village is accompanied by a request:

- "We've not had light for three days. Can you help with a generator?"
- "The roof is leaking again."
- "We need money for Uncle's burial."
- "They're increasing school fees again. Please do something."
- "You should come home for the festival and bring clothes for everyone."

No one asks how you're managing your city bills. No one cares if your salary hasn't been paid. No one thinks about your own family or children. You are the "city person." You have money.

A deep cultural code in many communities says, "Those who leave must give." This code is especially strong in villages. If you're the only one who's moved out or the most educated in your extended family, you're seen as the bridge, the connector between poverty and progress. While there is honor in that role, it is often accompanied by guilt, pressure, and unrealistic expectations. You're expected to:

- Bring gifts during every visit.
- Build the family home.
- Sponsor village projects.
- Give generously during weddings, funerals, and naming ceremonies.
- Sponsor one of your siblings to a world-class education.
- House and feed one of your family members.
- Stay humble and never complain—because "it's your duty."

It's a silent contract that city dwellers and diasporans never signed but are constantly bound by. Samuel moved from a small village in Benue State to Abuja after securing a government job. His family was overjoyed. He was the first to leave home and secure a job in an office. Every salary came with a list: money for firewood, school shoes, and cement for the family house. When he visited a year later, he brought bags of rice, clothes, and cash. But his mood shifted when people asked for more, and he declined. Some relatives started murmuring. His uncle asked, "Is it now that you've become a big man you've forgotten us?" That night, Samuel overheard someone say, "See him walking around like a politician. Let's see if his money will save him from death." He realized then that his success had become dangerous, not because of jealousy but because entitlement had turned affection into quiet resentment.

Priya grew up in a rural town in India. She moved to Mumbai to study at a university and stayed after landing a decent job. She sent money to her parents each month, covered her younger brother's tuition, and paid for a cousin's hospital bill. When she missed a family gathering during Diwali due to work deadlines, her aunt said on a family WhatsApp group:

"These city girls think they're too good for us now. What's all the education for if you can't respect tradition?" No one defended her. Not even her mother. Priya felt invisible. Her sacrifices had become expected, not appreciated. Her presence had become optional as long as the money kept coming.

In my research, urban migration often fuels cycles of dependency, whereas it should be a bridge to opportunity, not a trap of dependency. But when expectations aren't checked, it becomes a cycle:

1. Someone leaves the village for a city.
2. They work hard to survive and grow.
3. Family and community place unspoken financial duties on them.

4. They give and give while delaying their growth.
5. Their giving sustains the village, but at a personal cost. When they say "no" or set limits, they're branded as selfish, proud, or "lost to city life."

BREAKING FREE OF CITY SYNDROME

You can be true to your heritage without being overpowered by it. You can serve your people without being consumed by it. **This is how:**

1. You are not here to be the system; you are here to be the support system.
2. Set boundaries early: Inform individuals about what you can and cannot do. Do not wait until you are burnt out.
3. Normalize the absence of 'no', especially where you cannot offer. What happens is burnout.
4. Encourage independence: Encourage family members to work, acquire skills, or start their businesses.

MY TAKE

The city-village entitlement dynamic is subtle but powerful. It creates guilt where there should be pride, turns celebration into silence, and can twist love into a sense of obligation. But you have the right to rewrite the narrative. You can give, but you still have boundaries. You can love and still say no. You can honor your village and still protect your peace. Genuine respect is not about how much you give; it's about being valued for who you are.

Chapter 6

THE DARK SIDE OF ENTITLEMENT

Entitlement does not stop in personal relationships. If unresolved, it grows in magnitude, soon becoming hazardous and destructive. It penetrates families, communities, institutions, and countries. If not restrained, it becomes a cancer in the culture that infects every aspect of society, where dishonesty is tolerated, resentment is the norm, and integrity is absent.

However, the underlying proposition is that others should be depended upon to accomplish one's responsibilities. It is the attitude that other people are duty-bound to fix, fund, or address the problems one faces in life. Where such assumptions are prevalent, they create an environment where manipulation is encouraged, deception is justified, and corruption is tolerated. What began as a survival instinct became a weapon. People started to take advantage of one another, not because they were inherently evil, but because they had been conditioned to believe that taking was their right.

In many developing countries, this entitlement culture has contributed to poverty, underdevelopment, and broken trust cycles. People in positions of power often steal public funds without remorse, believing it is their turn to do so. Communities usually disregard leaders who loot, as long as the money benefits their people. Similarly, families shame their relatives abroad for not sending money, even if those relatives are barely surviving. This is the dark side. Entitlement turns gratitude into greed, makes exploitation seem like normal behavior, and destroys relationships that could have been built on partnership and purpose.

Today, people have lost their life savings because they trusted a family member to build a house abroad, only to find out it was all a lie. Others sent money to fund a small business for a relative, only for the relative to disappear once the money landed. Some have opened bank accounts for their cousins, only to discover months later that they have been used for fraudulent activity.

One Canadian sent nearly $20,000 to his younger brother in Ghana, hoping to invest in real estate and start a phone accessories business, receiving phone pictures, receipts, and voice notes. But he returned home after twelve months with no business and no property. Just his cousin, who had changed his number and could no longer be contacted. He felt betrayed not only for the lost money but also for the years of loyalty, love, and trust lost through an act of fraud.

These are not anomalous exceptions. These are typical,

normal, and even hallowed in some circles. When confronted, the culpable individual will probably say things such as:

- You are not suffering over there.
- "At Least I Did Not Steal from a Stranger"
- "We're family, you'll see."

But this is not understanding. It is manipulation in the guise of family, an indicator of a deeper issue or attitude problem. The entitlement mindset convinces people that their need justifies their behavior, that their suffering gives them the right to exploit, and that their connection to you is a license to take. But need never justifies greed. Poverty is never an excuse for theft; love does not absolve one from accountability.

The same entitlement within families often exists in political systems as well. Leaders steal from public coffers and fund lavish private lifestyles, while the masses struggle to access clean water and stable electricity. And yet, they are celebrated in their hometowns as "sons of the soil." Why? Because they brought home a piece of the pie, and the crime benefited them. And so the cycle continues, entitlement dressed in patriotism.

But there is nothing patriotic about corruption. And there is nothing honorable about betraying trust, whether in the name of family, community, or survival. Until we address the entitlement mindset, we will continue to build systems on shaky foundations. We will continue to lose good people to bitterness, good relationships to disappointment, and good intentions to betrayal. Entitlement robs us of empathy. It makes us forget that the one giving might be sacrificing, the one helping might be hurting, and the one supporting might be struggling. It reduces people to resources and relationships to transactions.

And it works both ways. Some diaspora individuals feel that they are owed obedience because they are perceived as benefactors. They use their assistance to oppress, manipulate, or gag their nationals. Their funding is their throne, upon which they sit in power. This is dangerous as well. Power is power, no matter the guise; no matter how it is misused, it causes deep pain.

This is not giving but taking in disguise. It becomes an exercise in domination masquerading as care. There is no need for the giver to yell because his cheque does it for him. "All that I have given for you" is the song. "Don't you forget who paid your tuition" is a leash. Bit by bit, inexorably, the dynamic of the relationship shifts from love to passive compliance.

In many homes, parents and siblings tiptoe around the diaspora provider, afraid to speak their truth, afraid to disappoint the one who holds the keys. The people who should feel safe, free, and loved begin to feel like employees, constantly trying to please the boss who signs the checks. The diasporan would take any opportunity to throw insults and make those he helps feel small; they look at their old classmates as lazy beggars and ignore their pleasantries. Take the case of David, who is Nigerian but lives in the U.K. He used to send remittances regularly to his brother Jacob, who is pursuing medicine. Along with each remittance came advice: "Don't socialize with those people." "Don't waste my money." "Don't fall in love with someone I would not approve of." Jacob initially obeyed. But over time, the weight fell on his head like a chain.

Years passed, but the relationship had strained after Jacob graduated and started working for himself. Resentment had already taken hold. He was grateful, but no longer trusted his brother. And David had no idea why things had come between them.

He had not even known his money came with strings attached. Then there's Aunty Rose, well-known among her extended family in Togo. She had lived in New Jersey for almost 30 years and was the primary provider for her siblings' children. She handled every school fee, every Christmas celebration, and every family need. But she also expected everyone to report to her before making decisions.

When her niece wanted to marry a man Aunty Rose disapproved of, she threatened to pull her support. "If you do this, don't count on me again," she warned. That marriage never happened. Her niece lives at home, still single, afraid to cross the woman who controls her life from another continent.

These are not just stories of giving gone wrong; they are stories of control, love turned into power, and generosity weaponized to maintain authority. Sometimes, the control is subtle. It's in the way someone brings up their contributions at every opportunity. It's in the looks exchanged when a decision is made without consulting "the one abroad." It's in the silence that follows defiance. It's based on the belief that money holds a voice in everything. It is weaponizing cash and gifts, using them as a tool to buy loyalty and submission.

But this isn't generosity; it's transactional charity wrapped in emotional strings. It must be said that ***you can support someone and still allow them to live their life.*** You can give without expecting control. You can love without puppeteering. The moment your help becomes a tool to dictate behavior, it ceases to be help. It becomes manipulation. True giving is rooted in respect. It doesn't silence. It doesn't control. It doesn't demand a seat at every table. It trusts. It empowers. It releases.

And for those who give from abroad, especially to parents, siblings, and younger family members, this is a call to check your heart. ***Ask yourself:*** *Am I helping to uplift or to control? Am I supporting, or am I silently ruling? Do I trust the people I give to or use my money to keep them dependent on me?* Because sometimes, the cycle of entitlement is a two-sided dance.

One side asks, and the other gives, then uses that giving as a crown. Both sides end up bruised. Both sides end up distant. And no one wins. The solution is honesty, transparency, and rebuilding relationships where giving flows from love, not leverage. Where people are free to grow, make mistakes, choose differently, and still be loved. Where giving is not a performance for power but an extension of grace.

Ultimately, no matter how far you've gone or how much you've given, *people deserve the dignity of being themselves.* Love doesn't dominate. Real generosity liberates. The antidote to entitlement is humility, mutual respect, and accountability. It shifts from "You owe me" to "How can we grow together?" From "Because of you, I didn't succeed" to "What can I learn from you?" From "I deserve" to "I appreciate."

This cultural shift doesn't happen overnight. It takes honesty, unlearning, and individuals deciding to break the cycle by setting boundaries, speaking up, and refusing to be manipulated by guilt. It also takes courageous conversations in families, faith communities, and political spaces. It takes leaders willing to model integrity. And it takes ordinary people, like you and me, choosing to live differently.

If we don't change our mindset, money will never be enough, giving will never be satisfied, and the relationship will never be whole. We must shine a light on this dark side, not to shame but to heal, not to judge but to transform. We cannot build a bridge between home and abroad if one side constantly drains the other. Healthy support is rooted in partnership, not pressure. It is rooted in respect, not revenge. It is rooted in accountability, not assumption. Until we embrace this truth, we will continue to mistake manipulation for love and lose good people to the quiet heartbreak of being taken for granted.

Entitlement, when left unaddressed, can develop into an emotional debt culture. People begin to feel owed, not because of an agreement but because of perceived suffering. The logic becomes twisted: "I am in need; therefore, you should fix it." What begins as hope turns into expectation, and what began as expectation becomes resentment.

The most painful of all is the emotional erosion of trust. People who once trusted one another with messages now question every communication. A harmless "hello" from a family member is stressful. "What do they want now?" is the first thought. It's a tiresome way to live, constantly looking to protect your heart while defending your wallet.

This is the reality for so many diaspora people. They are weary of giving and of being used. They are weary of being defined by what they can offer. They are tired of their goodness being misconstrued as weakness. They are weary of their relationships having been founded on what they could do, not who they are.

Even more devastating is how entitlement steals the

opportunity for real growth. Instead of encouraging innovation or self-reliance, people learn to wait, expect, and depend. Why learn a trade when your brother in Europe sends money every month? Why start small when your sister in the U.S. can fund your dreams in one transfer? Why try when someone else is struggling on your behalf?

But here's the truth that many don't want to face: You cannot outsource transformation. No amount of remittance can replace self-discipline, work ethic, or personal responsibility. Money can support, but it cannot sustain. And those who build their lives solely on the sacrifices of others will eventually face collapse.

A woman in Texas once shared how she financed her brother's move from Kenya. She covered everything: visa, plane ticket, rent for his first six months, and even helped him get a job. Within a year, he lost his job due to absenteeism, spent recklessly, and dared to ask for more money because "things are hard."

When she refused, he accused her of wanting to see him suffer. He moved out, blocked her, and told the family she abandoned him. That woman said she'd never felt so betrayed, not because of the money lost but because her love was weaponized against her. Before helping people back home, ask if they need your assistance. Not every dog that barks needs help; some are just barking so they can bite you.

This is the part of entitlement people don't discuss: the grief. The grief of losing people who once mattered, the grief of realizing your love was not enough, the grief of pouring into people who had holes they refused to patch, the grief of being needed but never truly known.

On the other hand, entitlement can also weaken communities. It robs them of the collective power that comes from shared effort. Instead of pooling ideas and talents to build something meaningful, people rely on one or two "blessed" individuals to carry everyone's burdens. It creates lazy systems. Communities that could thrive instead become dependent hubs, waiting for handouts from those who have "gone ahead."

Entitlement is evident in churches and mosques, where spiritual leaders often guilt-trip members into giving to the community, especially those abroad, while failing to teach sustainable empowerment. It appears in schools where students do not take their education seriously because they believe an aunt or uncle overseas will "connect" them to a job later.

It is evident in politics, where leaders often fail to invest in infrastructure because they expect the diaspora to "give back" to their country. It shows up quietly, persistently, and painfully in homes where expectations replace affection, and support is no longer a blessing but a burden.

Entitlement at its worst is emotional blackmail. Sometimes, it does not just state, "You owe me money." It states, "You no longer love us." It is cold shoulders at the family reunion sometimes. It is spiritual manipulation sometimes: "If you were committed before God, you would not take your people for granted." It turns charitable givers into emotionally trapped providers who are forever balancing on the tightrope of duty and self-protection.

But we must call it what it is, not to condemn but to be free. We must leave the cycle and say, "No more." No more giving from fear, being emotionally blackmailed, sacrificing your peace at the altar of expectation. Because here's what the dark side of entitlement fears most: boundaries. Boundaries are the antidote. Boundaries return dignity. Boundaries force people to grow.

They teach responsibility. They say, "I love you enough not to enable you." That kind of love is powerful because it invites maturity. Setting boundaries doesn't mean you stop giving. It means you stop bleeding. It means you give from wisdom, not compulsion. From a full heart, not an empty tank. It means you're choosing health for both you and them.

We also need to normalize accountability. It is alright to speak up if somebody betrays your trust, to call out if a family member lies, and to step back if someone is manipulating you. Silence only emboldens the cycle. The way forward is not anger but clarity. It is not retribution but healing. Healing is not possible without the truth.

The truth is that unchecked entitlement can be destructive. It destroys love, unity, growth, and trust. But when exposed, challenged, and redefined, it loses its grip. We don't have to keep repeating the same generational patterns. We don't have to keep building fractured relationships on fractured foundations. We can start again with clear boundaries, honesty, and respect. We can teach our families that sometimes love isn't loud, sometimes love speaks "no," and sometimes love holds back, not because we're arrogant but because we're smart. Above all, we can remember that we are not bad people. After all, we require balance, not selfish people.

We need harmony, and not horrible people, because we need space. We are human beings doing our best to love freely, live freely, and give out of the little we have.

Chapter 7

BRIDGING THE GAP AND CREATING SOLUTIONS

Following the stories, the heartache, and the silent sacrifices, there must be more than bare survival. There must be healing. The purpose of this book was not to demonize or shame but to reclaim the truth so that we may move forward with compassion and honesty.

This chapter is about moving forward, building bridges that do not buckle under the weight of entitlement but bear respect, boundaries, love, and growth in both directions. Closing the gap starts with reframing what help is. It means moving away from the idea that help is always financial and that real empowerment is collaboration, not dependence.

We must transition from a system in which one person helps all to one in which everyone contributes. The first step in this shift is learning to set healthy boundaries without guilt. For many in the diaspora, saying no feels like a betrayal. It feels like letting your people down. But a boundary is not rejection;

it's clarity. It's not selfishness; it's sustainability.

You cannot help others if you are emotionally, financially, or mentally drained. Boundaries are how we protect love from becoming resentful.

You don't have to justify every no. You don't need a PowerPoint presentation to explain your limits. It's okay to say, "I can't right now," and leave it there. The people who love you will understand or learn to. And those who withdraw because you've stopped overextending were never in a relationship with you, only with what you provided.

At the same time, those receiving support must begin to unlearn dependency and reimagine responsibility. The idea that help is owed simply because of a relationship must be replaced with a culture of gratitude, effort, and growth. When someone gives, it should be appreciated, not taken for granted. Help should be respected, not taken for granted, when offered.

Families need to have difficult yet healing conversations. We need to stop sweeping tension under the rug or using culture and religion to silence boundaries. Parents need to stop guilting their children into a lifetime of sacrifice. Siblings need to stop assuming that success means responsibility. Children need to be taught from a young age that effort, not entitlement, builds a lasting legacy.

We also need to begin encouraging self-sufficiency and entrepreneurship, not as a way of pushing people away but as a form of empowerment. Imagine a village where young people are taught how to turn their skills into small businesses, rather than waiting for someone in the city to send them money. Imagine families discussing investments, not just immediate needs. Imagine communities that value effort just as much as support.

That starts with practical investment, not just in money but also in mindsets. Sometimes, the greatest thing you can do for someone is not give them a monthly stipend, but rather provide them with access to education, skills training, or mentorship. It helps them build a resume, practice a business pitch, or connect them with someone who can offer advice.

We should redefine success in our homes as well. Success is not just about who sends the most money or builds the biggest house. It's about who has peace, is emotionally healthy, breaks generational cycles, and raises children who understand the value of things, not just wealth. We must also begin to honor those things.

Closing the gap also means learning to communicate expectations directly and honestly. Most breakdowns happen because nothing is ever said directly. Assumptions lead to disappointment. What we must do instead is be brave enough to have us talk like this:

- "This is what I can commit to giving each month, and no more."

- "I'm willing to invest in your business, but only after negotiating a proper plan and responsibility."
- "I love you, but I am not responsible for the family."
- "I'm willing to help your dreams come true, but I need you to meet me halfway."

These are difficult discussions and may anger people, but they are necessary. Without honesty, resentment lurks in the shadows, and quiet anger inevitably overflows into shattered relationships. We also need to build sustainable support systems, especially within the diaspora. No one should carry the weight alone.

Imagine a network where five people agree to collaborate on a single family project. Or a community fund where contributions are rotated. Or a mentorship program where diaspora professionals empower local youth, not with handouts, but with fundamental skills and lasting relationships.

Bridging the gap also means confronting the systems that reinforce entitlement. Religious institutions must teach stewardship, not just sacrifice. Cultural leaders must stop glamorizing wealth without discussing its underlying processes. Elders must start blessing efforts, not just results.

For those who have experienced betrayal, manipulation, or deep disappointment, bridging the gap may start with forgiveness—not for the other person but for yourself.

Forgiveness is the process of releasing the pain, allowing you to move forward without the weight of bitterness. It's saying, "That happened, but I will no longer live in the shadow of it." Reconciliation is sometimes an option, but it isn't always available. Either way, healing remains in your control. From that healing, you can build new habits, wiser giving, more explicit expectations, richer relationships, and stronger communities.

Finally, the goal isn't to stop giving. The goal is to give generously out of strength, not guilt, support without being consumed, and love without losing yourself. We don't need to burn the bridge between "abroad" and "back home."

We must fortify it with truth, honor, balance, and boundaries. This is how we begin to turn the story around and fix what has been broken by entitlement. That's how we bridge expectations with wisdom, clarity, and love.

Chapter 8

A NEW WAY FORWARD

There is always one point in each process at which we must decide whether to continue following our current path or branch off in another direction. This chapter is at such a point. Following page after page of story, pain, realization, and contemplation, A New Way Forward is a chapter of possibility and what healing, truth, and hope can build in the landscape of broken trust and burdened relations.

This is not a chapter of blame. This is a chapter of choice. And one of the best things about choice is that whether you live overseas, back home, in the village, or the city, it's yours. The point of departure along this new road is to choose clarity rather than assuming. No resentment unspoken. Time for an honest talk.

Families must begin talking about expectations before they become wounds. *It's okay to ask: What are you hoping for from me? What are your limits? How can we support one another in a way that doesn't leave anyone empty?* We must move away from unspoken pressure and toward mutual understanding. This means not only communicating more effectively but also listening more attentively. Sometimes, those abroad need to understand the perspective of those back home. And sometimes, those at home need to understand the exhausting reality of life abroad. A bridge requires both sides to walk toward each other.

Another key part of the new way forward is teaching financial responsibility and investment, rather than enabling short-term survival. We need to replace the habit of sending endless handouts with the mindset of building something that lasts.

Imagine if, instead of sending $200 a month indefinitely, you invested $1,000 in a relative's skills, paid for a course, bought them the necessary tools, or helped them register a legitimate business. And what if, instead of just passively waiting for freebies, they put effort into developing anything from what they were already given? That's how we become creators, not consumers.

We must also redefine generosity. Help does not always have to be financial. It can be time, wisdom, access, encouragement, and mentorship. Sometimes, the most powerful help you can give someone is believing in them enough to say, "You can do this. I'll walk with you, but won't carry everything for you."

In the same way, we must redefine what success looks like. Achievement isn't what you accumulate. It's what you build sustainably. It's what you protect in terms of relationships and what you hold on to in terms of peace. What is left after you're gone isn't what you contributed but what you instilled. As part of a new direction, it allows the next generation to think differently. We must teach children abroad and at home that they are not entitled to other people's labor, that love is not a transaction, and that mutual respect is more important than material exchange.

We must raise daughters who don't see marriage to someone in the diaspora as the ultimate escape plan. We must raise sons who don't see their sisters' or brothers' success as their inheritance. We must model healthy, balanced, and respectful relationships so that entitlement becomes the exception, not the expectation.

For those in the diaspora, this new way forward is about reclaiming your peace and identity. You can be proud of your success, rest, create boundaries, stop giving when it no longer feels good, and live your life fully without guilt. At the same time, you are invited to give from a place of purpose. Not because you're pressured to, but because you choose to.

You are powerful not because you fund other people's lives but because you can empower them to build their own. This is not just about saying "no" anymore. It is saying "yes" to what matters most: trusting relationships and support for what affirms dignity and the future, rather than the continuation of the past.

We must also make room for those who have been hurt by the concept of entitlement. They have been deeply wounded. What occurred to them is genuine, and what they feel is authentic as well. They can still heal. It may start with forgiveness, distance, or rewriting a new relationship on healthier terms. The road is different for everyone, but it's worth taking.

A new way forward also means building collective strategies. What if communities abroad formed giving circles —groups of five or ten people who agree to pool their resources and support causes back home with accountability, strategy, and transparency? What if we created mentorship networks instead of just remittance lines? What if diaspora support were structured, measured, monitored, and built around growth, rather than guilt?

That's how we begin to build bridges that don't collapse. Bridges are built not on one person's back but on shared strength, insight, and values. It's time to build relationships rich in love, not expectations. Relationships that aren't defined by how much you send but by how much you show up—in wisdom, in truth, in presence. Because love without honesty isn't love.

Support without sustainability isn't support. Giving without peace is a slow form of emotional erosion. But it doesn't have to stay that way. You can choose a different path, you can lead a different conversation, you can model a new kind of relationship and bridge the gap. In doing so, you can inspire a cultural shift, one conversation, one boundary, one brave "no," and one empowered "yes" at a time.

WHAT'S THE WAY FORWARD?

The current situation in most home countries has put their people at a critical crossroads. The government's inability to address problems and the lack of openness from individuals at home have left those in the diaspora desperate. Individuals in the diaspora have thus resorted to other measures focused on family welfare and personal lives.

Mutual trust is urgently needed to fill the gaps between individuals at home and individuals in the diaspora. Individuals at home need to realize that individuals abroad also have families and obligations. Those back home can demonstrate understanding and concern for their well-being through understanding and concern.

One of the best ways to gain trust is to be honest and open. It is easy to do without spending any money, yet rich dividends can be reaped in the long term. When those back home explain problems honestly and openly, they create an atmosphere of understanding and trust. This also strengthens the bond of understanding between diasporans and those who remain back home.

Individuals can hold the government accountable by

being truthful and open. When those in governance realize that what they do and decide is watched by those in the diaspora and back home, they remain responsible for their actions and strive to implement pragmatic solutions. Such a sense of responsibility can bring about positive change and reform within the government.

In contrast, failure to maintain and promote transparency, as well as to build and sustain trust, has disastrous implications. Insofar as governments of homelands do not pay heed to the interests and concerns of diasporas, they can only promote complete disengagement of the diaspora.

Complete disengagement of the diaspora is disastrous for the home country, as the diaspora plays a crucial role in facilitating the economic development of the homeland. However, if the diaspora feels neglected and marginalized, they can divert effort and capital elsewhere, depriving the homeland of badly needed financial inflows.

The diaspora can also prove to be an effective source of good change and development for the home nation. They acquire valuable skills, knowledge, and experience while living abroad. A trusting bond having been created with the diaspora, individuals back home can gain from this valuable resource and learn from their experience. This can lead to new ideas, financial prosperity, and well-being for the home nation.

For such genuine trust to be created, individuals at home

not only need to sympathize with them but also interact with, hear from, and include them in decision-making processes. They can follow them periodically, solicit their opinions and views, and integrate them into decision-making platforms. When consulted and valued, the diaspora will likely feel a sense of belonging and duty to its nation.

The current standoff between those abroad and those back home needs to be resolved immediately to build genuine trust. Those back home must recognize the families and obligations of those abroad and empathize by showing genuine concern for them. Sincerity and openness take center stage in building trust and holding those in authority accountable.

In other cases, those abroad tune out entirely, and their home country reaps the bitter harvest of resentment. In contrast, by listening and trusting overseas, those at home can use their positive input and move toward wholesome reform and development.

QUESTIONNAIRE

The chapter contains testimonies shared by people with similar experiences in their own words!

RESPONDENT A (NIGERIAN)

1. Have you experienced similar frustrations? If so, please explain.

I sent my dad's driver money for his daughter, who made it to university, and now he expects me to support him with every year's school fees - when I say I cannot, he dares to be upset with me. My uncle has poorly managed my grandfather's estate. All but one of the beneficiaries live overseas. After proposing options that would help make it more profitable and stabilize growth for all parties involved, my uncle insists on maintaining the status quo so he can benefit.

Every interaction is a transaction. When I'm looking for food, tailored clothing, or to change foreign currency, they say they know someone who can help, and then work with that person to agree on how they can get a piece of the action when it's all said and done.

I helped a schoolmate with some funds, and then several former classmates called me, lining up to ask for my help; I couldn't help them.

2. Do the folks back home recognize and appreciate the sacrifices you make? Please elaborate on your answer.

Folks back home think that because I live in the US, I somehow have unlimited access to funding, so they get upset when I tell them I cannot help because I don't have the funds to give.

3. How does the lack of appreciation for your sacrifices impact you? Please explain.

The lack of appreciation makes it more difficult to nurture the relationship due to increased tension resulting from their unwillingness to understand or accept your situation.

4. What do you believe are the key factors contributing to conflicts between you and your extended family back home?

The Unwillingness to break away from the every-man-for-himself mentality that corruption is built upon. There is a need to appear high-status by disregarding the less fortunate and adhering rigidly to cultural expectations. Traditional bureaucratic leadership limits power and decision-making to the familiar few. The view that Westerners are superior to us is why we have yet to enter our industrial age.

5. Are there any other key points you would like to add? Please explain.

The major challenge is people's unwillingness to practice

delayed gratification for the greater good. There is also a high standard for exterior appearances, but an extremely low standard for maintenance, upkeep, and continuous improvement. I hope this helps. A live conversation would do more justice, so please let me know if you'd like to discuss this further, and I'll be glad to make myself available.

RESPONDENT B (UGANDAN)

1. Have you experienced similar frustrations? If so, please explain.

Yes, I experienced the same issue this past Tuesday. I had recruited 10 boys to dig holes for a fence, and upon discussing the price of each hole, five boys wanted 1,000 shillings for each hole, and the other five boys wanted 500 shillings. I ended up using other people who asked for 350 shillings.

2. Do the folks back home recognize and appreciate the sacrifices you make? Please elaborate on your answer.

I have realized that the elderly and mothers appreciate the sacrifice, but the younger adults want to take advantage of every opportunity. Some younger men want everything to pass through them, and they complain about one another.

3. How does the lack of appreciation for your sacrifices impact you? Please explain.

The lack of appreciation demoralizes me. I have been working daily to support the mission, and when I contacted the community, I found that they had more complaints than appreciation.

4. What do you believe are the key factors contributing to conflicts between you and your extended family back home?

They lack confidence in things being given to them and have a lack of trust. People think others are receiving more stuff than they are. So, they become envious.

5. Are there any other key points you would like to add? Please explain.

There is the issue of promising to do work and failing to fulfill the promise. The job you would accomplish in a week can take a month and have many demands. This leads to stress and sometimes ends the vision.

RESPONDENT C (NIGERIAN)

1. Have you experienced similar frustrations? If so, please explain.

I have experienced this on a small scale, but I have heard stories from colleagues and friends about their frustrating experiences with this matter.

2. Do the folks back home recognize and appreciate the sacrifices you make? Please elaborate on your answer.

This can be answered with a yes. Folks back home appreciate what assistance we render them (i.e., financially). The unstable situations at home, including transportation, education, rent, food, and clothing, defy all known economic theories. This situation was not unprovoked, unintentional, or blindsided; it was man-made.

We have leaders with no foresight, empathy, direction, or plan, and woefully inadequate education in a 21st-century country where a high school certification officially qualifies you to run and hold a government position. A stable economy that allows individuals to plan and manage funds effectively really helps. A constant change and unpredictable economy will always disrupt and defy any known economic theories.

All the money sent home will never be enough with the constant uncertainties. I believe they appreciate the help provided, but since there are no other avenues to make ends meet, they ask for more from their relatives in the Diaspora.

3. How does the lack of appreciation for your sacrifices impact you? Please explain.

As mentioned earlier, in my way of thinking, I do not see it as a lack of appreciation, but the pressure of the circumstances they find themselves in drives them to ask for more, making us feel they are insensitive to our contributions. The economy is in a dire state, marked by instability, inflation, and a lack of direction.

The government's claim that the youth are lazy is far from the truth. The government must create an environment where citizens can thrive with ideas, create jobs, and find avenues to succeed. I never had issues sending funds home. My frustration was that they waited too long to ask.

When help is needed, I prefer that they let me know beforehand so I can plan accordingly. Springing up with surprise requests irritates me, and I usually ignore them.

4. What do you believe are the key factors contributing to conflicts between you and your extended family back home?

As stated earlier, pressure, panic, and hunger can cause one to think and act irrationally. So, I do not blame the folks back home for not understanding that life in the outside world is not a bed of roses as they think. You see, preaching to a hungry person about the difficulties you are experiencing in a civilized economy (in Europe, Asia, or the Americas) will fall on deaf ears. Our folks know that no matter what, you are in a better situation. Yes, you have bills to pay, your family to feed, school fees for your kids, rent, and you're working three jobs here to make ends meet, etc., and you're still in a much better situation. So, if we understand this and see how they think, we should avoid creating or acknowledging conflict because HUNGER will never let us think rationally.

5. Are there any other key points you would like to add? Please explain.

After somehow advocating for our folks back home, I am not blind to the fact that they also share considerable blame for the deteriorating economy and support of failed leaders and government. There are countless stories of people in the diaspora with good intentions and ideas of channeling the economy back home to benefit all citizens.

Several return to try to establish businesses that fail due to unstable economic situations, a lack of essential amenities and utilities, and other negative factors. Others believe that creating an effective economy requires a stable and knowledgeable government. When they return home to run for office, they are often frustrated by those already in place.

A more frustrating aspect is the people (our own) who intend to fight for and play a more crucial role in derailing these diaspora members' intentions. Their messages and ideas fall on deaf ears; they are lied to, and their funds are siphoned off.

The government is reaping, stealing, and amassing the nation's wealth for personal use. The gap between the haves and the have-nots is constantly widening every day. Opportunists in the government pocket the profits from the country's natural resources.

Roads are bad, schools and education are lagging and lacking adequate funds for support, hospitals and the health industry are deteriorating, farmers and farming have slowed down due to violence, and citizens are kidnapped for ransom. There is hunger everywhere.

There is also one thing I know and believe: when you are pushed too far and backed against the wall, you react and fight back. When will my country fight back? Who are we waiting for? We are too greedy, selfish, and self-centered, so we want to eat our cake and have it too. No one wants to sacrifice for another.

The situation back home is not unique to our country alone. Other countries we see thriving today experienced the same conditions we are in now. It is not new that the people of those countries have fought and shed blood for the changes and niceties they are enjoying now.

Older generations in those countries fought and shed blood for their rights and equality, which is what their younger generations are currently enjoying. An equal playing field where everyone in their country will partake in what their country produces.

RESPONDENT D (NIGERIAN)

1. Have you experienced similar frustrations? If so, please explain.

This situation sounds familiar, and I can relate. The only difference is that they appreciate my sacrifices, but each time I don't or can't help, they're quick to show their disappointment as if I haven't helped in the past. I wish they had not become entitled.

After helping a family member on numerous occasions with their kids' school fees, house rent, hospital bills, and more, he once told me that I'm a cruel person, just because, on one occasion, I couldn't provide the help he needed from me. He made me realize that all the help I've rendered in the past is nothing unless I help him until the end. My question then was, where and when is the end?

Remember, I said I've helped this individual numerous times, and he keeps asking for more. At some point, I began to think that the problem was me, not him, because I felt I had supported him for so long that it now felt like an obligation or entitlement.

I promised another family member I would support her monthly until her situation changed. I recall a specific instance where I sent her money that exceeded the current minimum wage in Nigeria, only for her to call me two weeks later to report that she had run out of groceries. I can go on and on about different situations.

2. Do the folks back home recognize and appreciate the sacrifices you make? Please elaborate on your answer.

Folks back home appreciate the sacrifices I make to support them. The only thing is, they become entitled, and when I can't help, they demonstrate some kind of disappointment, as if the support I rendered in the past doesn't count. That hurts because I sometimes go out of my way to help.

3. How does the lack of appreciation for your sacrifices impact you? Please explain.

As I mentioned earlier, when I'm unable to help, they often show a lack of appreciation for the assistance I've provided in the past. This hurts a lot because I sometimes go out of my way to help, just because I understand the economic situation back home is terrible. To them, though, you should be able to help every time because you live in better conditions abroad, which is not always true.

4. What do you believe are the key factors contributing to conflicts between you and your extended family back home?

The primary factor contributing to the conflicts is their lack of understanding that living or working in a developed country doesn't mean there's money flying everywhere for everybody to catch. We still work very hard to earn every dollar.

5. Are there any other key points you would like to add? Please explain.

The only thing I can add to the points mentioned above is that there are many unrealistic expectations of people who have traveled abroad, and little is not enough. You have to keep giving until you break. We also need to start explaining to the people back home that life is not easy here, too, and we work very hard to make those sacrifices we make for them, no matter how little.

RESPONDENT E (SPAIN)

1. Have you experienced similar frustrations? If so, please explain.

Yes, I have had the same experience. The fact is that siblings and friends in Nigeria see us all as sacrificial lambs for them. Most families believe it is acceptable to expect things from their siblings and friends residing abroad. Families and friends back home all think we have a magic money tree where we just go and have it. The biggest victims are our female siblings who have traveled abroad. Therefore, due to their soft-heartedness, families at home take advantage of their magnanimity for their self-centered and selfish interests.

2. Do the families back home recognize and appreciate the sacrifice made? Please elaborate on your answer.

Very few of them recognize and appreciate the support that is given. They all think we are millionaires once we step foot on the soil abroad, and that money-making is fun. Despite all your attempts to make them understand the contrary, it keeps falling on deaf ears. Again, families see it as a must that relatives abroad must continually help them at all costs.

Greed and a sense of entitlement have seeped deeply into their fabric. These and others are why many people abroad hesitate to share their contact information with relatives and friends at home. I know of people who pulled down their Facebook pages via Facebook Messenger due to excessive disturbances related to financial demands.

3. How does the lack of appreciation for your sacrifices impact you? Please explain.

As the saying goes, "Ingratitude is a zeal killer." Ingratitude fuels dissatisfaction, pushes people away from helping, and makes us miss the good. Gratitude can help us experience more positive emotions and improve positive relationships. There is a saying that a thankful heart will always earn more. I recall sending a substantial amount of money for a project years ago. Years later, the project was not realized, and when I asked for or sought an explanation, family members now took sides; some I sponsored through their university education, then stopped calling or checking on me. A friend once told me what he went through at the hands of his sibling. "He sent some money to his brother, but his brother rejected it, saying it was insufficient.

A few years later, he traveled, too. It has been over 10 years, and he has not been able to remit money home to anyone. That singular act of his has caused a rift between them ever since." While a simple "thank you" is an energy giver, a lack of appreciation will sap your energy and probably distort your emotions and thinking.

4. The key factors contributing to the conflict between my extended family back home and me are:

- Ingratitude
- Feeling of entitlement
- Expects a lot and is not willing to give back
- Laziness on the part of some family members
- Reaching out to you only when they are in need
- Insincerity
- Self-centeredness
- Incessant financial request
- Greediness
- Backstabbing

5. Are there any key points you would like to add? What I would like to add.

Without us here abroad, those ingrates (not all) will not die. Be very careful about who you help and how often you do it. Some time ago, I embarked on a building project back home. Could you believe that the person who defrauded me was a family member? Until I gave it to an outsider-a— friend—who took the project to its completion and constantly sent videos so I could periodically see its progress.

It is also worth noting that husbands and wives often find themselves at odds abroad due to the conflict caused by the constant demands from their families back home. Once they destroy your home, they will be the first to laugh at you. So, "a stitch in time saves nine."

While helping families and friends back home, do not do it at the expense of your immediate family. Help, but don't be foolish. Without you, they will not die. If you want, just fake your death (God forbid), and they will move on without you. Please always put yourself first.

RESPONDENT F (NIGERIAN)

I sympathize with you, my friend, on the one hand, and rejoice with you, on the other hand, that your experiences cannot compare with those of others who have suffered physical harm, even fatality. Dissecting the root cause- what, how, and why- is imperative to discuss this subject adequately.

First, a fundamental misunderstanding exists between these parties, who live in "two different worlds." One party has been given the impression that once you cross over the Atlantic Ocean or any of the borders," you have arrived at El Dorado of boundless riches.

That is the impression that those back home have been given. I said I have been given overtime, and they believe it. O'Guy will not tell them he is doing a menial job to survive. He will not tell them that the flashy car that Mukoro imported was stolen or bought on loan. Africans are known for being loud and love to show off.

The second reason is that the person left Africa with the El Dorado mindset, promising to return with help for their wards and neighbors, but never returned as promised or communicated the actual situation. Even if he tries, they will not believe him. So, to help, he has contributed to the Eldorado propaganda.

The root cause is multifaceted, with numerous contributing factors. Then, there is the desperation and hunger of Africans back home.

I do not recall attributing such experiences to my dealings or transactions across the Atlantic. So, what I am sharing is generally obtained from others. Maybe because, when I was on the other side of the Atlantic, I had my fair share of swindlers' exploitations, so I knew them and would not let down my guard. I have not let my guard down.

There is a saying, "The old man who walks up to a little girl to call her his wife is only inviting insult." The consequence is that activities are few in any environment of distrust; in other words, the cash flow rate is low, and the propensity for investment is also low.

RESPONDENT G (NIGERIAN)

1. Have you experienced similar frustrations? If so, please explain.

Yes, I have had similar experiences. I receive endless requests from people who think I have more than they do, and I often notice the sense of entitlement that arises because I live abroad. The current exchange rate also makes it easy for people to profile me as very rich, so some resort to exploitation when asked to help with tasks.

2. Do the folks back home recognize and appreciate the sacrifices you make? Please elaborate on your answer.

I'll say yes and no. Educating them on how things work abroad, the types of jobs available, and the sacrifices that people abroad make to make these things accessible has been beneficial for some. For others, the sense of entitlement remains high due to the reasons stated in response #1 above.

3. How does the lack of appreciation for your sacrifices impact you? Please explain.

It mainly has an emotional impact that leaves me feeling sad and disappointed. There should be some appreciation for helping others. Witnessing people's ungratefulness is quite painful and discouraging.

4. What do you believe are the key factors contributing to conflicts between you and your extended family back home?

Sense of entitlement, unrealistic expectations, dishonesty, and a lack of trust, among others.

5. Are there any other key points you would like to add? Please explain.

Those of us abroad can help those back home by educating them on the realities of life abroad. Working so hard and denying ourselves so much just to satisfy our longings will always leave people in deficit and could cause lifelong strain on their relationships.

RESPONDENT H (SENEGALES)

1. Have you experienced similar frustrations? If so, please explain.

Yes, I have experienced some of these same frustrations,

especially with the people in my own family, but I can only do what I can. Sometimes, it's hard to say no, but they have to understand that I also have my own family in the USA. They think that just because I've helped in the past, I should help every time.

2. Do the folks back home recognize and appreciate the sacrifices you make? Please elaborate on your answer.

My family back home appreciates the effort and sacrifices I make to help in any way I can, but they don't always understand when I'm unable to assist them. Sometimes, I'm unable to do everything they ask me to do, but they appreciate my efforts to help them. I must keep reminding them that living in the USA is not what it used to be. Things have gotten more expensive and even more competitive.

3. How does the lack of appreciation for your sacrifices impact you? Please explain.

Sometimes, it bothers me a little that I'm unable to meet their expectations. I know how it feels not to have, and even not be able to pay your bills, because I was in their shoes before coming to the USA. Therefore, not fully addressing their needs can affect me.

4. What do you believe are the key factors contributing to conflicts between you and your extended family back home?

It's not always about having enough money. They must understand that I have a family in the USA, too. I have a life to live, too. I am a big fan of helping others, but you must take care of your immediate family first. Then, what's left, you can certainly try to help others back home.

I remember fighting with my older brother, who used to live in France. My family back in Africa always asked me for money to help with something they needed to fix around their house. I called my older brother and told him he needed to help, too.

He then told my mother, and then my mother called me and got mad at me for getting angry at my brother. They think that just because I am in the USA, I should always be able to help.

5. Are there any other key points you would like to add? Please explain.

We must understand that we have made a huge sacrifice to be here. Since I came to the USA, I've only gone home to visit once. There was not enough time either, but I couldn't stay as long as I wanted because I couldn't be away from my family or job for that long. I hope my family back home understands that.

RESPONDENT I (NIGERIAN)

1. Have you experienced similar frustrations? If so, please explain.

It's not my family, but I heard a story of 2 individuals who lost a family member and had sent money home ahead of time. They gave their family members the task of preparing for the funeral before they arrived. When those individuals came for the funeral, the family had nothing ready. They used the money for their projects and assumed the individuals would just come with more since they came from America and must have had lots of money anyway.

2. Do the folks back home recognize and appreciate the sacrifices you make? Please elaborate on your answer.

There was an incident where I bought clothes that were not appreciated despite being expensive. The individual who received them looked down upon them and said they were not of quality and expected more. It felt demoralizing. This individual also believed that the amount of money given was insufficient.

When given a significantly larger amount of money than the original, they remained unsatisfied and did not appreciate the additional funds they received.

3. How does the lack of appreciation for your sacrifices impact you? Please explain.

We are all human beings and like to be appreciated. When appreciation is lacking, it can make one feel as though they should say, "Why bother to do more when the little I have done is not appreciated?"

4. What do you believe are the key factors contributing to conflicts between you and your extended family back home?

Extended families have caused conflicts because of greed.

5. Are there any other key points you would like to add? Please explain.

It can be frustrating because older people, who should know better, often use greed to divide others.

RESPONDENT J (NIGERIAN)

1. Have you experienced similar frustrations? If so, please explain.

I have had multiple experiences, but I will focus on the most recent ones. I wanted to set up one of my younger sisters in a farming business. She needed help purchasing a plot of land on the outskirts of Lagos; she came up with a proposal of 800,000 naira per plot, but she never mentioned that this was the price for an acre. After transferring the money, she returned with stories of an agency fee of 10% to the elders, another 10% to the local youths, and an additional 15% of the purchase price for the agent. When I insisted on getting my money back, the response was that I could not get a refund, but I was subjected to continuous extortion.

2. Do the folks back home recognize and appreciate your sacrifices? Please elaborate on your answer.

To be honest, while some relatives do appreciate it, the majority do not. Some enjoy it whenever you give or promise to give, but when you turn down a request, you are least appreciated or not appreciated at all, and you see and feel the silent treatment from the person you once turned down.

3. How does the lack of appreciation for your sacrifices impact you? Please explain.

How has the lack of appreciation affected me? Outside the fact that it has cost me unending resources, as stated or believed by one of my siblings, the Nigerian naira exchange to the dollar is in my favor. This notion is constantly rubbing it in my face, which gives a tremendous and ridiculous increase in their demands.

4. What do you believe are the key factors contributing to conflicts between you and your extended family back home?
Greed, extortion, and sometimes envy of reality and/or assumption of the reality of our existence or day-to-day living abroad. The assumption is that you earn or make enough to take care of the needs of everyone back home, allowing them to keep making unending demands that, in most cases, are above what you can afford.

A family member demands that I send him a vehicle as a settlement. Another relative with four kids and a living spouse still believes I must pay their children's school fees. It becomes a conflict when you attempt to turn down these unreasonable demands.

QUESTIONNAIRE SUMMARY

The influence of diasporans is pivotal in addressing issues at home or in their country of origin. By fostering trust, maintaining open communication, and engaging in collaborative efforts, diasporans and their relatives can contribute to a more prosperous and harmonious future. The government may be compelled to adopt a more responsive and responsible role driven by the collective efforts of its citizens, both at home and abroad.

SUGGESTIONS BY THE RESPONDENTS

Those of us abroad can help those back home by educating them on the realities of life abroad.

Respondent G. I do not blame the people back home for not understanding that life in the outside world is not a bed of roses, as they think.

Respondent C. People who think I have more than they do make endless requests and often notice the sense of entitlement that arises because I live abroad.

Addendum

THE ENTITLEMENT DIVIDE

Silent tension exists between people living abroad and their loved ones back home. It's not always loud or obvious, but it's there—a subtle emotional divide, a wall built by assumptions, silence, misunderstandings, and unmet expectations.

At the core of this divide is a lack of shared understanding. The person abroad often feels unseen, as if no one truly understands the pressure, the loneliness, and the sacrifices. Meanwhile, those back home feel forgotten, abandoned by the person they once knew, now changed by "foreign air."

This divide widens every time assumptions replace communication. Whenever someone says, "You're in America, what's the big deal?" The person abroad says nothing out of fear of being misunderstood. Every time, needs are dismissed, boundaries are ignored, and love feels like labor.

To bridge this entitlement divide, both sides need to step into each other's world physically and emotionally. Those back home need to listen to what is said and what has been hidden. Listen to the fatigue in the voice, the pressure behind the silence, and the courage it takes to say, "I'm not okay."

And those abroad must also make space to listen to the pain of poverty, the desperation behind some of the requests, and the dignity people are trying to preserve even as they ask for help.

Bridging the divide takes grace, willingness, and honesty. It takes both sides to learn that no one is better than the other; they are simply different. Struggling in Toronto is no better than struggling in Kampala. Struggling in London isn't more noble than struggling in Lagos. They have different paths, different pressures, and the same humanity.

This is how the divide begins to heal: through empathy, not assumption. Through truth, not guilt. Through connection, not comparison.

THE BURDEN OF SUCCESS

Success should be beautiful, light, joyful, and earned. However, for many in the diaspora, success is heavy, carries weight, comes with strings attached, and rarely belongs to them alone. This is the burden of success: You're not just living for yourself. You become the hope of the family, the savior of the community, and the walking evidence that "someone made it." Your win becomes a communal entitlement.

And it doesn't stop. The more you succeed, the more

you're expected to sustain everyone else. Promotion? Prepare for more requests. New house? Someone will ask for rent. Travel photos? Someone will ask why you haven't sent money lately. It becomes hard to celebrate yourself. Every blessing is met with a fresh demand. Every joy is followed by guilt. Some even learn to hide their success, downplaying their lives to avoid the flood of expectations.

The burden of success also isolates. You can't vent. You can't cry. You can't say, "I'm overwhelmed." Because people think you have it all together. And if you say you're tired, they call you ungrateful. But here's the truth: success doesn't mean you're limitless. It doesn't mean you owe everyone. It doesn't mean you are the solution to every family crisis. You can succeed and still need help. You can achieve and still rest. You can succeed and still say, "Not right now."

Releasing the burden means permitting yourself to live, enjoy, not explain, not rescue, and not shrink. Your success is not a crime, and your joy is not a betrayal. You've worked hard, and it's okay to protect the fruit of that work with wisdom. Success should never feel like punishment. It should be shared but never stolen and should be honored, not exploited.

NO LONGER THE ATM

One of the most dehumanizing aspects of entitlement is how it reduces people to roles: the ATM, the one who sends money, the one who fixes things, the one who must always give. Conversations fade. The connection disappears. What's left is a cold transaction: request, transfer, silence. But you are more than what you provide. You are a sibling. A daughter. A son. A friend. A human being. And somewhere along the way, we have to say: *enough*. Not from bitterness but from truth. From the desire to have relationships rooted in love, not obligation. No longer being the ATM means rewriting the dynamic.

It means giving with intention, not guilt. It means explaining your limits before resentment builds. It means inviting people into conversations, not just transfers. It means asking people, "What's your plan?" before you offer help. This shift will be uncomfortable. Some will resist. Some will be offended. But the ones who genuinely care about you will stay. They will grow. They will learn to relate to you again, not as their financial institution but as family.

You can still be generous. You can still be helpful. You can still be available. But you don't have to be a machine. You have the right to say no. You have the right to ask questions. You have the right to redefine the terms of your giving because love should never cost you your dignity.

Conclusion

A CALL FOR CHANGE

A quiet cry has been rising between expectation and exhaustion, between love and obligation. It comes from the man in a one-bedroom apartment sending half his paycheck back home, the woman holding her family together with silent sacrifices, and the child abroad trying to balance the weight of legacy with the freedom to live. That cry is this book.

Bridging Expectations was never just about entitlement. It's about the emotional cost of being needed but not seen. It's about the fractures that form when relationships are defined by how much you give, rather than how well you're loved. It's about healing the silent grief of those who have been guilted into silence and burdened into performance.

But more than anything, it's about hope. The hope is that conversations can change culture and that boundaries can repair relationships. When spoken with love, that truth can free both the giver and receiver. The work ahead will not be easy. It requires courage, unlearning, and accountability on both sides, those who give and those who receive.

To those in the diaspora: *You are not selfish for needing* *rest. You are not wicked for saying no. You are not ungrateful for choosing peace. Your presence matters as much as your provision. You deserve love that doesn't come with a price tag. You can redefine your relationships without losing them if you do it with grace and truth.*

To those back home: *Gratitude is a language that heals. Effort is a seed that multiplies. Love is not measured by how much someone spends but by how much they care. Releasing people from guilt is a crucial step in welcoming them back into the relationship. Learn to support the one who supports you. Grow with them. Build with them. Walk beside them, not behind them, with outstretched hands. Add value to their lives; you can also buy things for them and ship them to them without being paid or asked. You can help them with a skill you have, such as designing a website or creating a dress. It must be two-sided, and not one person always gives.*

Bridging the gap is not about abandoning culture. It's about evolving it. It's about holding on to the beauty of communal support while letting go of the chains of unspoken pressure. It's about rewriting the family narrative, not from a place of dependency, but from one of dignity, guilt, and grace.

We have the power to shift this culture, create new systems of giving and receiving, rebuild trust where it has been lost, and tell our children a better story—the one where support is mutual and success is celebrated without strings. Love is not weighed in money but in meaning. This is your invitation. To pause. To reflect. To choose a different way forward. Because the bridge between expectation and understanding has already been built with your honesty, healing, and hope. ***Now, it's time to walk across it. Together.***

www.ingramcontent.com/pod-product-compliance
Lightning Source LLC
Chambersburg PA
CBHW040927210326
41597CB00030B/5210